Pratim Biswas, Gregory Yablonsky (Eds.)

Aerosols

Also of interest

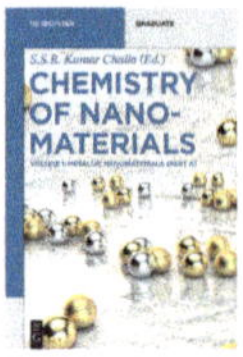

Chemistry of Nanomaterials. A Practical Guide
Volume 1 A Metallic Nanomaterials (Part A)
S.S.R. Kumar Challa (Ed.), 2019
ISBN 978-3-11-034003-7,
e-ISBN 978-3-11-034510-0

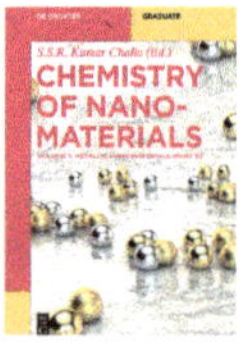

Chemistry of Nanomaterials. A Practical Guide
Volume 1 B Metallic Nanomaterials (Part B)
S.S.R. Kumar Challa (Ed.), 2019
ISBN 978-3-11-063660-4,
e-ISBN 978-3-11-063666-6

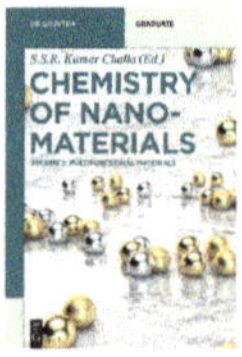

Chemistry of Nanomaterials. A Practical Guide
Volume 2 Multifunctional Materials
S.S.R. Kumar Challa (Ed.), 2020
ISBN 978-3-11-034491-2,
e-ISBN 978-3-11-034500-1

Nanoscience and Nanotechnology.
Advances and Developments in Nano-sized Materials
Marcel Van de Voorde, 2018
ISBN 978-3-11-054720-7,
e-ISBN 978-3-11-054722-1

Nanoanalytics.
Nanoobjects and Nanotechnologies in Analytical Chemistry
Sergei Shtykov (Ed.), 2018
ISBN 978-3-11-054006-2,
e-ISBN 978-3-11-054201-1

Aerosols

Science and Engineering

Edited by
Pratim Biswas and Gregory Yablonsky

DE GRUYTER

Editors
Prof. Pratim Biswas
Department of Chemical, Environmental and Materials Engineering
University of Miami
1251 Memorial Drive
Coral Gables 33124
United States of America
pbiswas@wustl.edu

Prof. Gregory Yablonsky
Department of Energy, Environmental and Chemical Engineering
Washington University in St. Louis
St. Louis 63130
United States of America
gyablons@slu.edu

ISBN 978-3-11-073096-8
e-ISBN (PDF) 978-3-11-072948-1
e-ISBN (EPUB) 978-3-11-072955-9

Library of Congress Control Number: 2022932617

Bibliographic information published by the Deutsche Nationalbibliothek
The Deutsche Nationalbibliothek lists this publication in the Deutsche Nationalbibliografie; detailed bibliographic data are available on the internet at http://dnb.dnb.de.

Cover image: Pattadis Walarput/iStock/Getty Images Plus
Typesetting: Integra Software Services Pvt. Ltd.
Printing and binding: CPI books GmbH, Leck

www.degruyter.com

Contents

List of contributing authors

Chapter 1
Introduction
Pratim Biswas
Department of Chemical, Environmental and Materials Engineering
University of Miami, Coral Gables
FL 33146, USA
AND
Gregory Yablonsky
Department of Energy, Environmental and Chemical Engineering
Washington University in St. Louis, MO 63130, USA

Chapter 2
Early stages of particle formation in aerosol reactors: measurement and theory
Yang Wang
Department of Energy, Environmental and Chemical Engineering
Washington University
St. Louis, MO 63130
USA
AND
Department of Civil
Architectural and Environmental Engineering
Missouri University of Science and Technology
Rolla, MO 65409
USA

Chapter 3
Aerosol methodologies for synthesis of materials
Xiang He
Department of Mechanical and Nuclear Engineering
Virginia Commonwealth University
800 E. Leigh St., Richmond, VA 23219
USA
Wei-Ning Wang
Department of Mechanical and Nuclear Engineering
Virginia Commonwealth University
800 E. Leigh St., Richmond, VA 23219
USA
E-mail: wnwang@vcu.edu

Chapter 4
Calibration and applications of low-cost particle sensors: a review of recent advances
Jiayu Li
Department of Bioproducts and Biosystems Engineering
University of Minnesota
Twin Cities
1390 Eckles Ave, St. Paul, MN 55108
USA
Pratim Biswas
Department of Chemical, Environmental and Materials Engineering
University of Miami
Miami, FL 33146
USA

Chapter 5
Carbon dioxide conversion methodologies
Liang-Yi Lin
Institute of Environmental Engineering
National Yang Ming Chiao Tung University
Hsinchu 300
Taiwan
E-mail: lylin@nctu.edu.tw

Chapter 6
Aerosol science and nanoscale engineering enabling agriculture
Ramesh Raliya
Department of Energy, Environmental and Chemical Engineering
Washington University in Saint Louis
MO 63130
USA
AND
IFFCO – Nano Biotechnology Research Center
Gandhinagar, Gujarat – 382423
India
Email – rameshraliya@iffco.in

https://doi.org/10.1515/9783110729481-203

Katie Halwachs
Department of Energy, Environmental and Chemical Engineering
Washington University in Saint Louis
MO 63130
USA

Chapter 7
Airborne transmission of SARS CoV-2: variants and effect of vaccines

Pratim Biswas
Department of Chemical, Environmental and Materials Engineering
University of Miami
Miami, FL 33146
USA

Sukrant Dhawan
Department of Chemical, Environmental and Materials Engineering
University of Miami
Miami, FL 33146
USA

Pratim Biswas and Gregory Yablonsky

Chapter 1 Introduction

Abstract: The Chapter defines the scope of this edited book and defines the field of aerosol science and technology in a broad sense. The chapter also provides an Introduction to the entire book and outlines the layout of the various following Chapters.

Keywords: Aerosols, Nanoparticle Technology, Aerosol Science and Technology, Global Challlenges

This book describes some of the applications of the wonder discipline of aerosol science and technology. Clearly, the word "aerosol" often conjures many different meanings – starting from stuff that comes out of a spray can, to something that now causes infectious diseases such as COVID. However, there is a better understanding, and here is a generalized definition of the term "aerosol." The word has a Greek origin and refers to particles suspended in air or a gaseous medium. Many generic representations are often made – ranging from systems of small particles, smoke, haze, fog, dust, soot, mist, nanoparticles, airborne viruses, and bacteria [1–5].

The definition of aerosol containing the term "particle" is very interesting in the first place, as it is typically constituted of a stable cluster of molecules. The suspended particle is in a gaseous medium also consisting of molecules that are not in a cluster state. The interactions of the suspended particle with the gaseous molecules are ubiquitous and result in rather interesting physical and chemical behavior. Particles can be in a range of sizes, the smallest being the molecular dimension, to orders of magnitude larger entities. Often, we restrict ourselves to the upper size range in the 100-micrometer scale, as larger particles would always have a short, suspended lifetime. While we mention the interaction of our particles with gaseous molecules, we must mention that particles in vacuum have also been of interest – in many syntheses reactor systems; and of late as dust in lunar environments. These particles behave as in the free molecular regime and are of great relevance for practical applications.

There are other important characteristics of aerosols. A few of these include the shape of the particles and the chemical composition. Often particles are made of

Pratim Biswas, Department of Chemical, Environmental and Materials Engineering, University of Miami, Coral Gables, FL 33146, USA
Gregory Yablonsky, Department of Energy, Environmental and Chemical Engineering, Washington University in St. Louis, MO 63130, USA

https://doi.org/10.1515/9783110729481-001

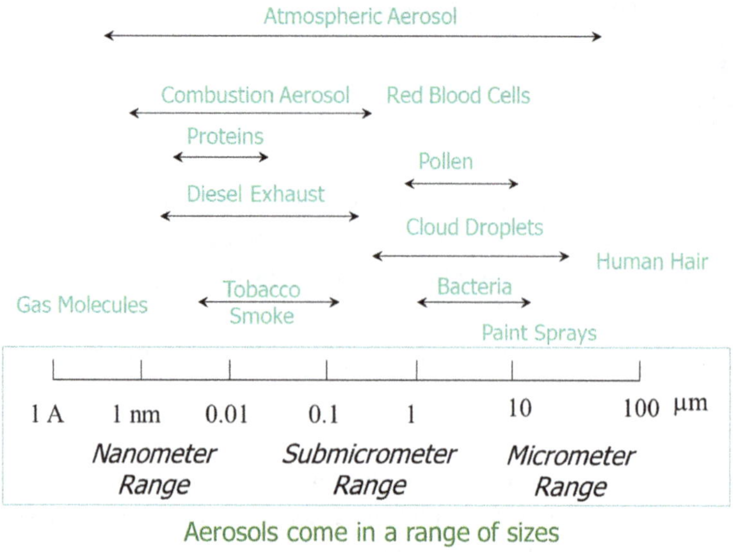

Figure 1.1: Different size ranges and examples of aerosols.

multiple chemical species, and in addition to the size (often defined as an equivalent diameter), the shape and chemical composition are also described. The particles can also be agglomerates or aggregates of primary particles, and this characteristic is also an important feature. A hierarchy of aerosol structures is created by the variety of intermolecular forces between the molecules of particles (1) Ion – dipole; (2) H-bond; (3) dipole-dipole; (4) ion-induced dipole;(5) dipole-induced dipole; (6) dispersion (London forces)with energies (1) 40–600 kJ /mol; (2) 10–40 kJ/mol; (3) 5–25 kJ/mol; (4) 3–15 kJ/mol; (5) 2–10 kJ/mol; (6) 0.05–40 kJ/mol, respectively

In aerosol systems, there are different levels of hierarchy, an aerosol particle, and an ensemble of particles. The ensemble is characterized by the special configuration of particles.

Consequently, the unique feature of aerosol systems transport, both diffusional and convective, is a two-scale characteristic that is important. The first scale is related to the single-particle movement. The second one corresponds to the transport of particle agglomerates, and these two scales are coupled depending on particle size and properties, and process parameters as well. Transport of aerosols exhibits specific features under the influence of external fields, thermal, gravitational, and electromagnetic. Generally, aerosols must be characterized as temporal and spatial “transport-reaction” systems with *dynamic functioning*. Many processes of these systems, i.e., coagulation/agglomeration/restructuration, deposition process via the “aerosol-surface” interaction, etc., have different characteristic times [6–13].

The fundamental understanding of aerosol systems allows us to better understand both engineered and natural systems. Figure 1.2 is illustrative of the fundamental

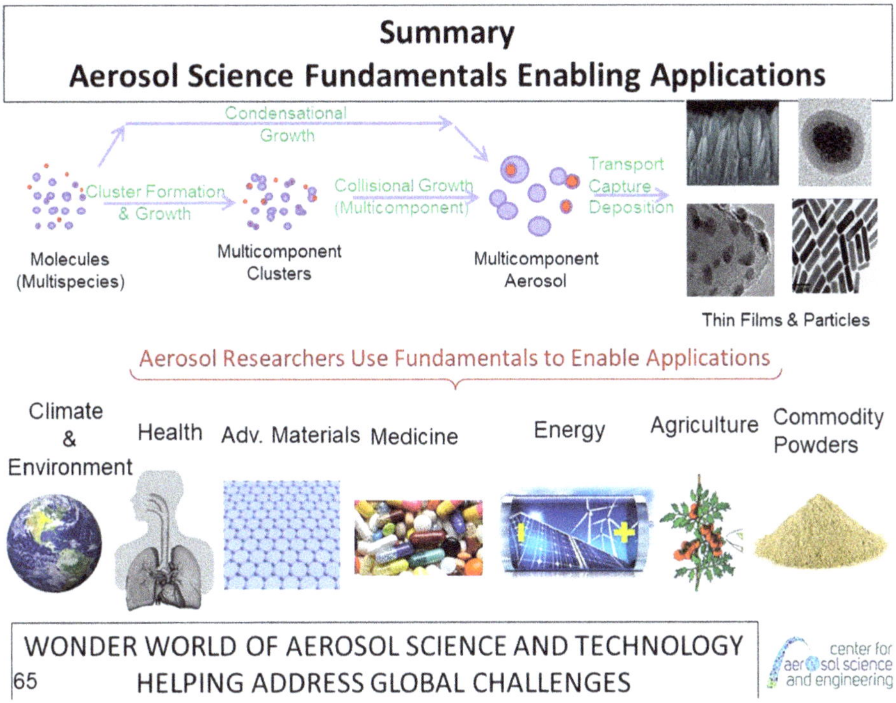

Figure 1.2: Fundamental understanding of aerosol science enables applications in a broad set of areas.

processes, and the variety of applications. In fact, the field of aerosol science and technology is often referred to as an "enabling discipline," due to the uniqueness of the field, and applicability in promoting understanding and applications for the societal benefit [5].

This monograph is a compilation of several areas of application of the wonder world of aerosol science and technology. Chapter 2 by Dr. Yang Wang focuses on developing an understanding of the early stages of particle formation. Many advances have been made; however, even today some of the theories to predict particle formation, especially in high-temperature systems, and in complex multi-species systems remain inadequate. In contrast, advances have been made in the measurement of small clusters, and a bridge has been made between molecular species determination (by spectroscopy) and the early stages of particle formation. This is discussed in detail by Dr. Yang Wang.

Chapter 3 is a summary by Xiang He and Dr. Weining Wang of various aerosol reactor methodologies for the synthesis of a variety of different materials. They provide a comprehensive review of the subject. It should be noted that aerosol science and engineering has been empirically used for about a century in the chemical industry to produce materials. With the advent of better tools and understanding, a variety of materials can now be synthesized. The advantage of aerosol reactors is that they can be scaled up, and the understanding over the years has aided in these developments.

Chapter 4 by Dr. Jiayu Li is a nice review of miniaturized particulate matter (PM) sensors that are now very pervasive. This chapter elegantly points out the need for calibration and delves into various uses of these sensors in a distributed manner. PM sensors such as the ones described will enable applications not only in domains such as smart cities, but also help in the safe transport and habitation of humans of extraterrestrial bodies such as the moon and Mars.

Dr. Liang-Yi Lin describes orderly the methodologies that will be needed to overcome the carbon dioxide issue and the importance of not only mitigating emissions but converting them to useful products. The use of aerosol science and technology to produce novel catalysts to enable this challenging problem is described in Chapter 5.

Smart and precision agriculture are important methodologies that have to be developed to feed the ever-burgeoning population of the earth. And this must be done in an environmentally safe manner. Dr. Ramesh Raliya and Katie Halwachs describe the importance of nanotechnology in helping accomplish these goals in Chapter 6 of the book.

Chapter 7 addresses a very topical area of airborne transmission of infectious agents, and a comprehensive model to predict the resultant risks of viruses. The coupling of aerosol transport behavior with inhalation exposure and resultant risk due to infection are integrated into a comprehensive model. Dr. Pratim Biswas and Dr. Sukrant Dhawan present it in this chapter and highlight the importance of such approaches in guiding public health decision-making.

"There's Plenty of Room at the Bottom." Richard Feynman said these words establishing the nanoscience. It is remarkable that aerosol *nano*clusters govern the hierarchy of the multi-level structures bridging both *micro-* and *macro*-world, and *the Universe*. In that sense, aerosol science and technology is truly an enabling discipline, and this book elegantly highlights some of the various application areas.

References

[1] Friedlander, S.K. (2000). Smoke, Dust, and Haze. Fundamentals of Aerosols Dynamics, 2nd edition, Oxford University Press, New York, Oxford, 408.

[2] Seinfeld, J.H., Pandis, S.N. (2016). Atmospheric Chemistry and Physics: From Air Pollution to Climate Change, 3rd edition, Wiley.

[3] Flagan, R.C., Seinfeld, J.H. (1988). Fundamentals of Air Pollution Engineering, originally published by, Prentice Hall, now available through Caltech Library.

[4] Hinds, W.H. (1999). Aerosol Technology, Wiley.

[5] Biswas, P., Wang, Y. (2019). The Wonder World of Aerosol Science and Technology: Problem Sets with Solutions, Amazon Publishers.

[6] Zaveri, R.A., Easter, R.C., Fast, J.D., Peters, L.K. (2008). Model for simulating aerosol interactions and chemistry. *Journal of Geophysical Research* 113(D13204):1–29.
[7] Zaveri, R.A., Shilling, J.E., Jiumeng Liu, A.Z., Bell, D.M., D' Ambro, E.L., Gaston, C.J., Thornton, J.A., Laskin, A., Lin, P., Wilson, J., Easter, R.C., Wang, J., Bertram, A.K., Martin, S.T., Seinfeld, J.H., Worsnop, D.R. (2018). Growth kinetics and size distribution dynamics of viscous secondary organic aerosol. *Environmental Science and Technology* 52:1193–1198.
[8] Murzin, D.Y. (2010). Kinetic analysis of cluster size dependent activity and selectivity. *Journal of Catalysis* 276:85–91.
[9] Murzin, D.Y. (2012). On cluster size dependent activity and selectivity in heterogeneous catalysis. *Catalysis Letters* 142:1278–1283.
[10] Murzin, D.Y. (2015). Cluster size dependent kinetics: Analysis of different reaction mechanisms. *Catalysis Letters* 145:1948–1954.
[11] Lee, M.H., Cho, K., Shah, A.P., Biswas, P. (2005). Nanostructured sorbents for capture of cadmium species in combustion environments. Environmental Science & Technology 39: 8481–8489.
[12] Biswas, P., Wu, C.Y. (2005). 2005 critical review: Nanoparticles and the environment. *Journal of the Air & Waste Management Association* 55:708–746.
[13] Snytnikov, V.N., Dudnikova, G.I., Gleaves, J.T., Nikitin, A.S., Parmon, V.N., Stoyanovsky, V.O., Vshivkov, V.A., Yablonsky, G.S., Zakharenko, V.S. (2002). Space chemical reactor of protoplanetary disk. *Advances in Space Research* 30(6):1461–1467.

Yang Wang

Chapter 2
Early stages of particle formation in aerosol reactors: measurement and theory

Abstract: One important frontier of aerosol science and technology is characterizing the initial stages of particle formation below 10 nm in size. Sub-10 nm particles play important roles by acting as seeds for particle growth, ultimately determining the final properties of the generated particles. Understanding the new particle formation in the atmosphere has recently become a heated topic in atmospheric science. In aerosol synthesis, tailoring the aerosol properties requires a thorough understanding and precise control of the particle formation processes, which in turn requires characterizing nanoparticle formation from the initial stages. This chapter reviews recently developed experimental and computational techniques for studying nascent particles in aerosol reactors. Using experimental and numerical data, it also summarizes the major research findings regarding particle formation and growth in aerosol reactors.

Keywords: aerosol reactor, aerosol measurement, formation mechanism, aerosol modeling, aerosol instrumentation

2.1 Introduction

Nanomaterials have received considerable attention during the past few decades, due to their particular mechanical, thermal, acoustic, electrical, and optical characteristics which result from their large specific surface area and quantum effects. Annually, over 2 million tons of functional nanoparticles, including titania, silica, and carbon black, are produced [83], and they have been widely employed in areas, including catalysis, solar energy utilization, paint production, sensor technology, and the rubber industry. Other ambitious uses of nanoparticles are also being projected, including water treatment using functional carbon-based nanoparticles and drug delivery using polymeric or metal nanoparticles. Compared to wet chemistry approaches (e.g., sol-gel), aerosol technology has proven advantageous for manufacture of commercial quantities of

Acknowledgments: The author would like to acknowledge Dr. Pratim Biswas for the advices and discussions and the financial supports from the National Science Foundation (2132655).

Yang Wang, Department of Energy, Environmental and Chemical Engineering, Washington University in St. Louis, St. Louis, MO 63130, USA; Department of Civil, Architectural and Environmental Engineering, Missouri University of Science and Technology, Rolla, MO 65409, USA

https://doi.org/10.1515/9783110729481-002

nanoparticles because of its high-throughput production, fast processing time, and facile process design [83, 89, 146]. Moreover, nanoparticles with complex compositions, such as most mixed oxide ceramics, can be manufactured with lower cost and less environmental harm by aerosol technology than by wet-chemistry [124].

On the other hand, particulate matter (PM) in atmospheric and aquatic systems poses threats to human health [196], air quality [3], and regional and global climate [144]. Fine combustion particulates, especially soot particles, are cytotoxic and can have substantial adverse effects on cardiovascular and pulmonary health [28, 100]. In addition, soot absorbs radiation strongly over the entire spectral regions and contributes significantly to global warming via direct radiative forcing. It also reduces the surface albedo by settling on snow and ice surfaces. PM emission can be categorized into primary and secondary types, depending on whether the PM was emitted directly from the source to the atmosphere or was generated in atmospheric reactions between contaminants [133]. In terms of primary PM emission, industrial and residential thermal systems and vehicle engines are the most important sources of $PM_{2.5}$ (PM with aerodynamic diameters smaller than 2.5 μm) [49, 85]. These thermal processes also generate large amounts of NO_x, SO_x, and volatile organic compounds (VOCs) that react to form fine particulates, which can further grow via vapor condensation and coagulation [118, 161, 177].

A substantial amount of functional nanomaterials and harmful airborne PM originates from aerosol reactors operated under high temperatures (100 ~ 2,000 °C). An aerosol reactor converts gas-phase precursors to solid nanoparticles in a single-step, using thermal energy generated from combustion or electric heating. Tailoring the properties of nanoparticles synthesized from aerosol reactors and controlling the PM generated from emission sources require a thorough understanding of the particle formation processes. However, currently, our knowledge of particle formation and evolution in aerosol reactors is significantly deficient. These deficiencies are attributable to a combination of factors, including (1) a lack of sensitive, accurate, and non-invasive techniques for measuring the physical and chemical characteristics of particles during their evolution in aerosol reactors, and (2) a lack of computational techniques to simulate the nucleation of the gas species that form particles and the detailed chemistry involved in particle growth. These deficiencies become more severe when studying particle evolution in the early stages, for example, when the particle size (D_p) is below 10 nm.

To understand early-stage particle formation and growth mechanisms, we must develop new techniques to measure particle size, morphology, fine structure, and elemental and molecular composition. Thus far, these efforts have been challenged by the distinctive properties of sub 10 nm particles, whose high diffusivity causes large sampling losses during measurement, and whose low charging probability limits the use of electrical methods for particle size classification and detection. Moreover, on this small length scale, the physical and chemical properties of these nascent particles start to strongly depend on each other, challenging the suitability of classical theories that describe the physical and chemical behaviors of particles. For example, theories describing the nucleation rate of particles have not quantitatively agreed with measured data, due to the complex interactions among reaction kinetics, molecular cluster surface properties, and charge involvement [30]. But more importantly, measurements of particle nucleation rate are compromised by the fact that current particle counters cannot detect particles below 3 nm efficiently, because these particle counters are highly sensitive to the chemical properties of the detected particles [81, 159]. In aerosol reactors, classical nucleation theory typically predicts a critical nucleus size that is within 1 nm in aerosol reactors [152], a size that cannot be easily detected by *in situ* methods. The *ex situ* measurement of these nascent particles becomes even more difficult due to the need to dilute the particle system and quench the reaction without affecting the chemical equilibrium.

Because experimentally characterizing particle formation and growth is difficult, modeling studies can play an important role in developing and improving experimental processes. Moreover, the reproducibility of modeling approaches helps to predict the reactor performance and to understand the importance of certain mechanisms by isolating their potential effects. With current modeling approaches, simulating the interactions among the chemical and physical properties of early-stage particles may become quite complex, because many conventional models monitor only the evolution of particle size.

Despite these challenges, new experimental techniques have been developed for measuring nascent particles. New modeling approaches across multi-length scales have also been developed. This chapter will highlight these new tools for measuring early-stage particles, along with their advantages and weaknesses. Using experimental and numerical data, it will summarize the major research findings regarding particle formation and growth in aerosol reactors.

2.2 Experimental techniques for studying early-stage particle formation

Particles generated from aerosol reactors can be analyzed by online and offline methods, depending on whether the analysis of the measured data is conducted simultaneously with (online) or separated from (offline) the operation of the aerosol reactor. Moreover, online measurement can be divided to *in situ* and *ex situ* types, depending on where the measurement is conducted and whether the measurement influences the aerosol formation process. For example, inserting an aerosol sampler in a flame aerosol reactor reduces the local temperature during combustion, potentially affecting the evolution of the aerosol properties. On the other hand, laser diagnostics have proven to be an effective *in situ* technique that does not interfere with the formation of particles in aerosol reactors. This section summarizes the major online (*in situ* and *ex situ*) and offline methods for studying early-stage particle formation in aerosol reactors.

2.2.1 *Ex situ* sampling and aerosol measurement

For *ex situ* measurement, particles are extracted from the aerosol reactor and transported to instruments for analysis. Here, we review the experimental setups, including particle sampling systems and instruments for *ex situ* particle analysis.

2.2.1.1 Particle sampling systems

Two types of dilution samplers are commonly used for measuring particles generated from high temperature aerosol reactors [46, 96], namely the hole-in-a-tube (HiaT) sampler [63, 195] and straight tube (ST) sampler [97, 155, 166] (Figure 2.1). These two types of dilution samplers have been evaluated previously for sampling soot particles and agglomerates generated in sooting flames and nanoparticle synthesizing flames [7, 8, 46, 195]. However, the sampling of early-stage particles ($D_p < 10$ nm) is more challenging, because the dilution samplers ideally should quench all particle growth dynamics and gas-phase reactions during the particle transport in the sampling line. Moreover, the flame perturbation caused by inserting the dilution probe should be minimized. Recent studies show that for sub 3 nm particles, particle coagulation can be effectively quenched when the dilution ratios of these two samplers are above certain values [165]. This effective quenching has also been verified by numerically comparing the different time scales describing particle coagulation, precursor reaction, and particle transport in the sampling system. Regarding the perturbation introduced by these sampling probes, one study conducted in a combustion system found that the perturbation was modest and did not qualitatively affect the particle size distribution, although the temperature was significantly decreased by the sampler's presence [23]. Another study, however, points out that the orifice sample flow can notably impact the local flow field, temperature, and particle residence time [131].

In addition to achieving sufficient dilution, the fully assembled sampling system should also be calibrated to evaluate the penetration efficiency. The sampling of nascent particles requires that particle coagulation and diffusion loss are minimized, which can be achieved by secondary dilution of the sampled aerosols and reducing the length of the sampling system. In addition, bends and elbow units in the sampling line significantly affect the penetration efficiency of particles, and direct calibration or simulation is required to assess nanoparticle penetration efficiencies for any flow system containing these units [164]. Metal tubes are recommended for sampling the incipient particles, because tubes made of organic materials can easily build up surface electric fields, reducing the penetration efficiency of charged

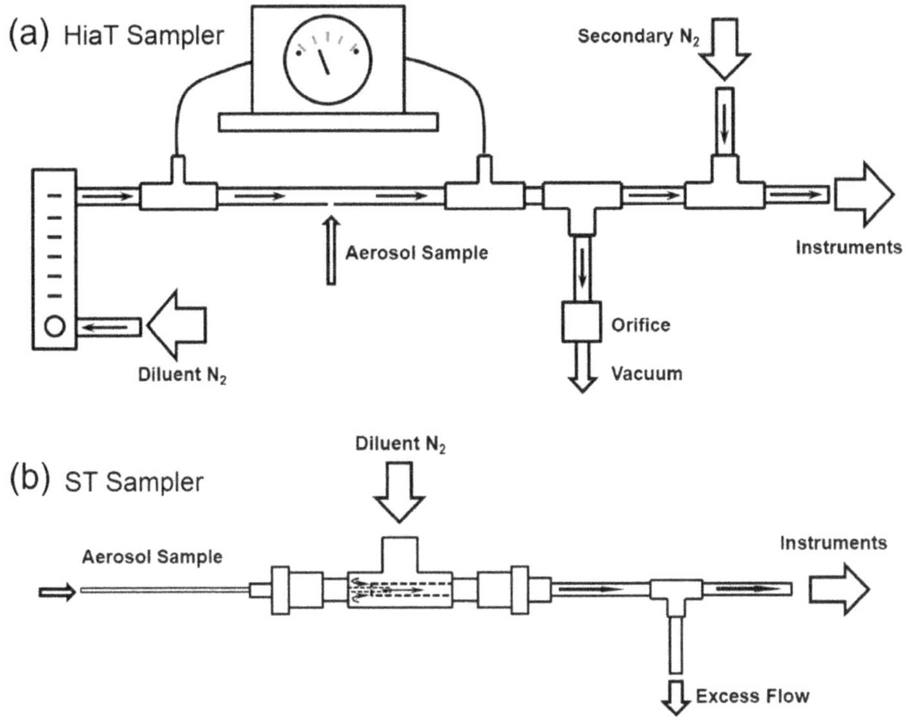

Figure 2.1: Common designs of dilution samplers for measuring aerosols at high temperatures: (a) hole-in-a-tube (HiaT) sampler and (b) straight tube (ST) sampler. (Figure reprinted with permission from Biswas et al. [8], Copyright 2018 Elsevier.).

particles [86]. Moreover, semi-volatile compounds may be emitted from the surface of these non-metal sampling tubes, strongly interfering with the measurement of particles smaller than 3 nm [150, 166].

2.2.1.2 Particle measurement techniques

Aerosols can be characterized by many aspects, among which their concentration, size, and mass are most important. Various other aerosol properties, such as their morphology, scattering coefficient, mixing state, and toxicity, can be derived from or closely related to these properties. The following section reviews both established and novel instruments for studying early-stage particles in aerosol reactors.

Particle concentration measurement

Particle concentration is a critical parameter in evaluating the impact of aerosols. The concentrations of submicron particles are normally measured with an aerosol electrometer (EM) or a condensation particle counter (CPC). An EM measures the current created by the collection of charged aerosols on a metal filter. It is ideal for the measurement of nascent charged particles because these particles normally carry only one charge, and the metal filter can reach almost 100% efficiency in collecting these charged particles. However, the electrometer has two drawbacks. One disadvantage is that an electrometer measures only charged particles, but the charging of nascent particles is highly dependent on their chemical composition [62]. Thus, the electrometer may capture only a glimpse of the entire aerosol ensemble, which may lead to considerably biased conclusions. The detection limit of an electrometer is another disadvantage. Presently, the lowest detection limit is in the range of fA (femtoampere, 10^{-15} A). For aerosols with a concentration of 1 cm^{-3}, in order to create a response above this detection limit, the flow rate of aerosol going through the electrometer must be at least 600 lpm (liters per minute), but current designs cannot handle such a high flow rate (the maximum flow rate is 10 lpm, for the TSI Electrometer, Model 3068B).

CPCs, on the other hand, have a very low detection limit (down to 0.1 cm^{-3}) and measure particle concentration regardless of particle charging state. A CPC grows a sampled particle by condensing the vapor of a working fluid on the surface of the particle, until it has an optically detectable size (~1 μm). The smallest particle that can be condensationally grown (or activated) is determined by the Kelvin equation, which considers the effects of the saturation ratio, surface tension, temperature, and molecular volume of the working fluid [36, 145]. The Kelvin equation dictates that the commonly used butanol-based CPCs have a smallest detectable size of 2.5 nm. The need to understand the transition from molecular clusters to particles has motivated the development of a number of condensation particle counters with detection limits approaching 1 nm. A detailed calculation using the Kelvin equation and data on more than 800 organic compounds suggests that diethylene glycol (DEG) is the best candidate working fluid for activating the condensational growth of sub 3 nm particles [55]. However, because of the low saturation vapor pressure of DEG, the grown particles are not large enough to be directly detected by the optical components of the CPC. Hence, a two-stage CPC was developed, where the first stage uses DEG to grow particles to a moderate size, and the second stage uses butanol to measure the concentration of the grown particles (Figure 2.2). Condensation with water vapor is also of interest, because differences in activation by water and organic vapors can reveal information about particle chemistry [62]. A three-stage "versatile" water condensation particle counter (vWCPC) was designed, whose cut-off size could be conveniently reduced to 1.6 nm by tuning the temperatures of the different stages [50]. These techniques greatly facilitate the observation of particle formation and growth in the initial stages.

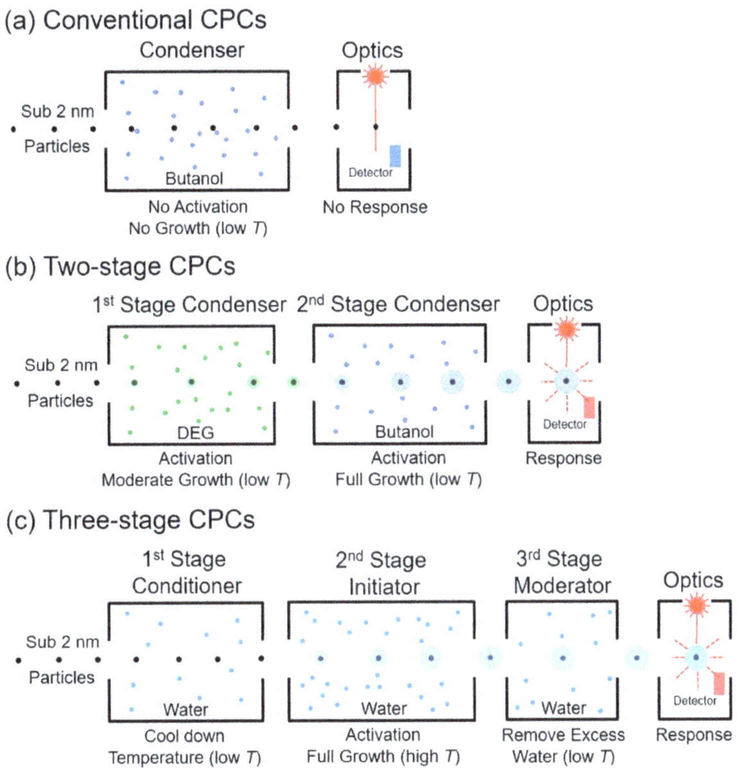

Figure 2.2: Working principles of a (a) single stage, (b) two-stage, and (c) three-stage CPC. (Figure (a) and (b) adapted with permission from 8, Copyright 2018 Elsevier].

Particle size measurement

Particle size can be defined by many metrics, such as physical size, volumetric size, mass-equivalent size, mobility size, and aerodynamic size. Because particles are seldom spherical, and because direct visual measurement of submicron particles is challenging, mobility size is the most widely used size metric. It is calculated from the electrical mobility (Z_p) of particles, using the Stokes-Millikan equation (Eq. 2.1), and is measured as the particle moves under electrical fields. In the Stokes-Millikan equation,

$$Z_p = n_e eC/3\pi\mu D_p \tag{2.1}$$

n_e is the number of charges carried by the particle, e is the elementary unit of charge, C is the Cunningham slip correction factor, and μ is the dynamic viscosity of the carrier gas. Differential mobility analyzers (DMA) and drift tubes are commonly used for measuring the mobility size of nascent particles.

A DMA acts as an electric filter, allowing only particles with certain electrical mobilities, i.e., mobility sizes, to pass through and be further counted by particle counters such as an EM or a CPC. The mobility size of the classified particles is dependent on the voltage applied on the DMA, the DMA flow condition, and the DMA geometry. DMAs with different geometries have been manufactured, for example, a cylindrical DMA [68], a parallel-plate DMA [132], and a radial DMA [189]. A cylindrical DMA, composed of two concentric metal cylinders, is the most widely used. Theoretical analyses using the stream functions of the flow and electric fields in a cylindrical DMA show that, the mobility (Z_p) of classified particles is a function of the voltage applied to the DMA (V); geometric parameters, such as the bullet length (L), bullet inner radius (R_1), bullet outer radius (R_2); and the flow rates, including sheath flow rate (Q_c) and aerosol flow rate (Q_m) [68]. Z_p is given by

$$Z_p = \frac{(Q_c + Q_m)}{4\pi LV} \ln(\frac{R_2}{R_1}) \tag{2.2}$$

When particle diffusion is neglected, the resolution of the mobility classification (the ratio of the transfer function peak mobility to the full width at half maximum of the transfer function) equals the ratio of the sheath flow rate to the aerosol flow rate [68]. Accordingly, the idea of the "transfer function" was developed, which describes the probability of particles with an arbitrary mobility being classified at the voltage corresponding to a specific mobility. The shape of the transfer function is triangular under the non-diffusing condition, whereas it becomes broadened when particles are diffusive, leading to a deviation in the particle trajectory. The diffusion broadening of the DMA transfer function is a significant issue when we measure sub 3 nm particles, since the diffusion coefficient is inversely proportional to the square of the particle diameter in the free molecular regime [36, 57, 186].

The direct method for reducing the broadening effect of DMA transfer functions is to reduce the residence time of particles in the DMA, which is done by enhancing the sheath flow rate. Based on this concept, much effort has been made to develop high-resolution DMAs that are optimized for sub 3 nm particle measurement [25, 130, 143]. The Herrmann DMA [51] and half-mini DMA [25] are two representative designs of high-resolution DMAs. Both DMAs can accommodate sheath flow rates higher than 800 lpm without developing turbulent flow, and the length of the classification zone is significantly reduced in order to decrease the particle residence time. A computational fluid dynamic simulation study shows that for a half-mini DMA, both the height and resolution of its transfer function are much better than those of other DMAs in sub 3 nm particle size range [186]. A comparison between the mobility diameter ($D_{p,Z}$) and volume diameter ($D_{p,v}$) of sub 3 nm nano-drops further shows that, when $D_{p,Z} > 1.5$ nm, $D_{p,Z}$ within 1.4% of $D_{p,v} + 0.3$ nm [79]. The deviation becomes larger for smaller particles because the cluster and the dipole it induces in the gas molecules interact. This relationship helps in evaluating the physical size of early-stage particles.

The particle size distribution $(n(D_p))$ can be further calculated based on the number concentration $(N(V))$ of particles classified under different DMA voltages. When particle multiple charging effect is considered, solving $n(D_p)$ is an ill-posed inverse problem, and the Twomey-Markowski algorithm is commonly used [99, 154, 170]. However, the multiple charging effect of early-stage particles is negligible, and $n(D_p)$ can be solved by the equation

$$n(D_p) = \frac{dN}{d\ln D_p} = \frac{\alpha N(V)}{\frac{Q_a}{Q_s}\eta_{DET}(D_p)f_c(D_p)\beta(1+\delta)} \tag{2.3}$$

where β and $\delta$are dimensionless flow parameters. The constant α can be derived from the relationship between Z_p and D_p. Detailed expressions can be found elsewhere [199]. $\eta_{DET}(D_p)$ is the detection efficiency of the particle counter, and $f_c(D_p)$ is the particle charging efficiency.

DMAs, as well as a number of other recently developed devices [142, 170, 188], can be categorized as spatial electrical mobility spectrometers, because these instruments separate particles by directing them along mobility-dependent trajectories. Because the residence time of the classified particles in the DMA is constant and independent of particle size, diffusional broadening leads to degradation of the instrument resolution at smaller particle sizes. A drift tube-ion mobility spectrometer (DT-IMS) for measuring aerosol particles in the range of 2 to 11 nm was constructed [113], and applied to study the structure and stability of organic nanoclusters [117]. In a DT, the charged particles, sampled at a specific time, migrate across an electrostatic gradient toward a detector, and the electrical mobility of a charged particle is inversely proportional to its transit time through the drift tube. The shortened residence time of smaller particles reduces the uncertainty of the particle movement, retaining a narrow transfer function for nascent particles.

Particle mobility size is important in that it directly governs the particle motion resulting from the external forces. However, being dependent on the particle interaction with the surrounding air, the mobility size is not purely intrinsic to the particles. For example, in eq. (2.1), a change of temperature and pressure may easily cause changes to the Cunningham slip correction factor and gas dynamic viscosity (temperature-dependent only). Besides, when the particles are not spherical, these properties depend on the particle orientation relative to their movement, and the interpretation and use of these properties often become complicated. Therefore, it would be more convenient if we could measure the intrinsic properties, among which particle mass is important. Moreover, for nascent particles at early stages, direct mass measurement also reveals the chemical composition of the particles. Instruments for measuring the mass of early-stage particles are introduced as follows.

Particle mass measurement

The aerosol particle mass analyzer (APM) [31, 32] and centrifugal particle mass analyzer (CPMA) [112–114] are commonly used for measuring the bulk mass of aerosol particles. Both instruments use the balance between the electrostatic force created by an electric field and the centrifugal force created by a rotating flow channel. Both instruments also consist of two rotating coaxial cylinders, with a voltage applied across the channel between them. While charged particles carried by sheath flow travel through the gap, the flow channel also rotates around the centerline of the cylinders, creating a centrifugal force that counteracts the electrostatic force. Only particles with a specific mass to charge ratio (m/z) are able to traverse the outlet slit and be measured by a particle counting device. The difference between the APM and CPMA is that the inner electrode of the APM rotates at a same angular velocity as the outer electrode, while the inner electrode of the CPMA rotates with an angular speed faster than the outer electrode of the CPMA. The settings in the CPMA create a stable system of forces that significantly improves the transfer function of the classifier [113]. Combining particle mass and mobility size measurements, one can conveniently obtain the particle density and morphology information [115]. The APM and CPMA typically have a lower detection limit on the order of 1 attogram, which is approximately 12.4 nm for a spherical particle with a bulk density of 1 g cm^{-3}. For even smaller particles, one needs to rely on a mass spectrometer to measure the actual mass of the particles.

The aerosol mass spectrometer (AMS, Aerodyne Inc.) is well established for measuring the chemical compositions of aerosols with aerodynamic sizes above 40 nm. However, to resolve the participating compounds during particle formation in the initial stages, the chemical composition measurements of atmospheric ions/clusters must be improved. The newly developed Atmospheric Pressure Interface Time-of-Flight Mass Spectrometer (APi-TOF, Tofwerk AG) can detect and measure ambient ions in a mass/charge range up to 2,000 Th [58]. It has a mass accuracy higher than 0.002% and a mass resolving power of 3,000 Th/Th.

2.2.2 *in situ* measurement

As introduced above, there are powerful *ex situ* tools for assessing particle size, morphology, and composition. However, these tools all require extractive sampling from the aerosol reactors, which cause significant perturbation to the processes under study. There is thus a need for non-intrusive *in situ* methods that can provide this type of information. Laser diagnostics have proven effective in probing the physical and chemical evolution of particles in aerosol reactors. The mechanism and application of various types of laser diagnostic techniques have been summarized in a number of reviews [20, 29, 103, 104]. This section introduces the working principles of laser-induced incandescence (LII) and laser-induced breakdown spectroscopy (LIBS), which are promising techniques for measuring early-stage particle formation in aerosol reactors.

LII is the most commonly used technique for measuring the distribution and volume-fraction of soot during combustion [157, 172]. In this application, a pulsed or continuous laser heats the particles to the sublimation point of around 4000 K. The resulting incandescence signal is approximately proportional to the volume fraction of soot particles. Detection by LII requires that the particles absorb at the laser wavelength strongly and are refractory, i.e., sufficiently mature and graphitic to sustain temperatures above ~3,500 K without vaporizing or undergoing photolysis. Other than soot particles [6], LII can also detect gas-borne titania, silicon, and metal nanoparticles generated from aerosol reactors [18, 21, 90, 140].

LIBS is a type of atomic emission spectroscopy which uses a highly energetic laser pulse to ablate, atomize, and excite samples in a laser-induced plasma. The plasma reveals the emission lines for each element in the plasma volume. Difficulties in analyzing nanoparticles using LIBS arise because all the gas and particle phase species are excited, making it difficult to differentiate the sources of the signals. Recently, a low-intensity phase selective-LIBS (PS-LIBS) technique (Figure 2.3) has been developed to study the initial stages of particle formation in aerosol reactors [175, 190, 193, 194]. With no macroscopic spark observed during the measurements, disturbances in the flow field are negligible, and no delay in the signal collection is needed. Given the reduced power, where the laser fluence is between the breakdown thresholds of the gas and particle phases, high selectivity is observed between emissions from atoms within nanoparticles versus atoms from gas-phase molecules [183, 191]. This selectivity of the low-intensity LIBS is advantageous for tracking nanoparticle formation and for measuring particle volume fractions during gas-phase synthesis or other processes.

2.2.3 Offline analysis

Nascent articles formed in aerosol reactors can also be collected and characterized chemically and physically by offline measurements. The sampling procedure is the crucial point, because the procedure should not falsify or change the particle properties, to ensure that the sampled material is representative of the particles in the particle formation environment. Under ambient conditions, aerosol particles can be facilely collected by a filtration system for offline analysis [54, 82]. Aerosol reactor-generated particles can be collected in a similar manner, but only after proper dilution (Section 2.1.1) due to their high temperature and high aerosol concentration [17, 52]. Another common way of collecting particles is thermophoretic sampling [48, 149, 151], and the design of a thermophoretic sampler for monitoring coal combustion is introduced by Gao et al. [41]. A cold surface placed in line with flow streamlines produces a strong temperature gradient, driving particles towards the surface by thermophoretic forces. To reduce flame perturbation, cool substrates can be inserted into aerosol reactors with a sampling time of few milliseconds. For both

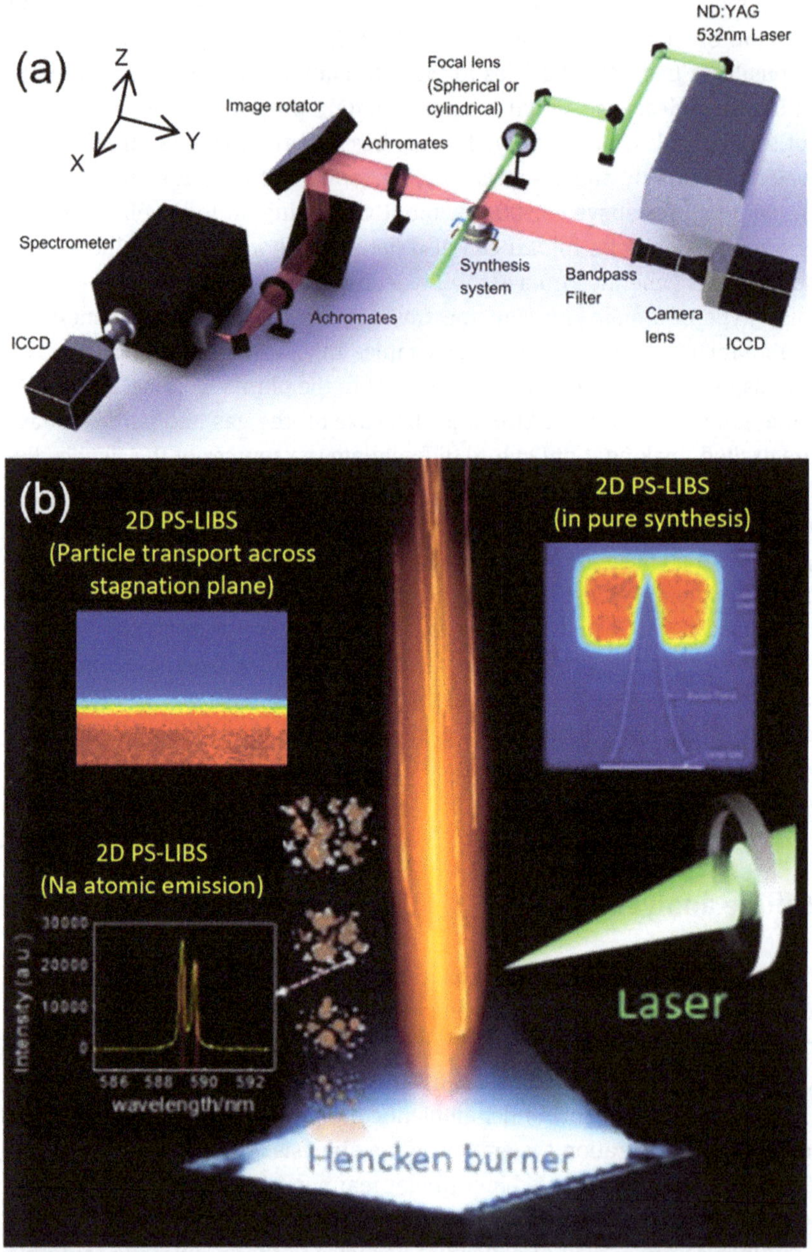

Figure 2.3: (a) Schematics of a typical PS-LIBS optical diagnostic system studying the particle formation in during flame synthesis (figure reprinted with permission from Zhang et al. [190], Copyright 2014 American Institute of Physics]. (b) Applications of PS-LIBS for studying particle formation during coal combustion and flame synthesis (figure reprinted with permission from Zhang et al. [193], Copyright 2017 Elsevier).

particle collection methods, the collected samples can undergo spectroscopic characterization and microscopy analysis. Size distribution functions and partial chemical composition information can be obtained by transmission electron microscopy (TEM) [53, 126], high-resolution transmission electron microscopy (HR-TEM) [129, 141] and atomic force microscopy (AFM) [5, 22, 106].

2.3 Measurement of early-stage particles in aerosol reactors

Different types of aerosol reactors have been designed for studying functional nanoparticle synthesis and PM emission. A more detailed classification of aerosol reactors based on the states of the precursor feed (in solid, liquid, or gas form) can be found elsewhere [146]. Depending on the source of the heat driving the particle formation and growth, these aerosol reactors can be categorized as the flame aerosol reactors, furnace aerosol reactors, glowing wire generators, spark discharge generators, and so on. Due to their wide usage, this section focuses on the experimental findings in flame and furnace aerosol reactors, while measurements in other types of aerosol reactors will be introduced briefly.

2.3.1 Flame aerosol reactors

A flame aerosol reactor utilizes the heat generated from gas or liquid fuel combustion to accomplish the precursor reaction, particle nucleation, and particle growth by condensation, coagulation, and sintering in a single step [162]. It is one of the most scalable and stable methods for synthesizing functional nanomaterials [83, 89] and producing soot nanoparticle standards for investigating detailed soot formation mechanisms [12]. Figure 2.4 shows the general particle formation and evolution processes during soot formation and combustion synthesis of functional nanoparticles. Although sharing similar particle dynamics, the final properties of particles generated from a sooting flame and a particle synthesizing flame are different due to the different flame conditions and chemical processes. Different types of flames can result from the way in which the fuel and oxidant are mixed in a burner and by their flow rates. As a result, flames can be further categorized as laminar premixed, laminar diffusion, turbulent premixed, or turbulent diffusion flames, which are studied in different scenarios. Studies investigating the particle characteristics in a sooting flame and a nanoparticle synthesizing flame are summarized as follows.

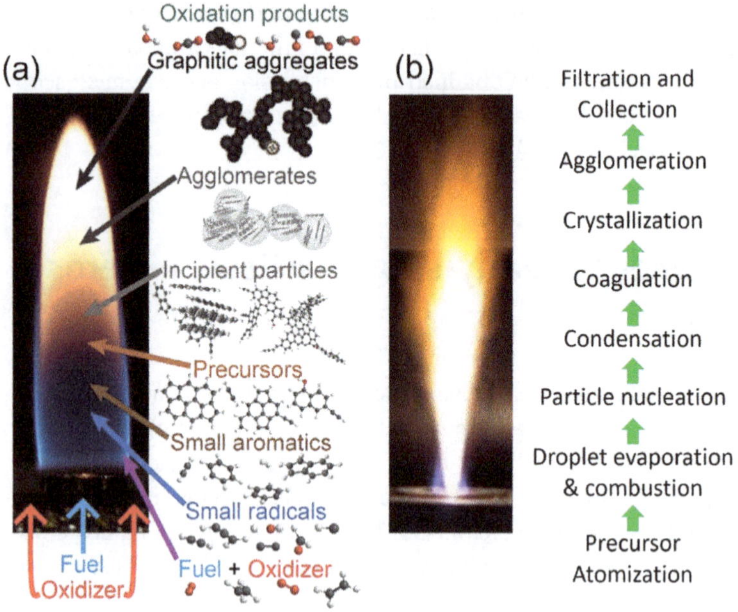

Figure 2.4: General particle formation and evolution during (a) soot formation in a diffusion flame and (b) combustion synthesis of nanoparticles in a flame spray pyrolysis reactor. (Figure (a) reprinted with permission from Michelsen [103], Copyright 2017 Elsevier; Figure (b) reprinted with permission from Phanichphant et al. [123], Copyright 2011 MDPI.).

2.3.1.1 Sooting flame

The experimental identification and characterization of soot particles in combustion zones are vital to our understanding of the mechanisms controlling particle formation, growth, and oxidation in combustion systems. To facilitate the study of soot formation, the flames are typically in the laminar regime. In order to clarify the mechanism of soot formation from the gas to particle phase, many studies have reported the measurement of early-stage particle size distributions using DMAs [11, 195] . The first endeavors probing sub 10 nm soot particles were made by Sgro et al. through a series of experiments using high-resolution DMAs to study the charging characteristics of nascent soot particles generated from ethylene flames [136, 137]. Nascent particle size distributions were measured under a series of equivalence ratios, and showed that the smallest particles observed in flames have a size of 1 nm, consistent with previous measurements using other techniques [19]. The charge distribution of the nascent soot particles was found to be very close to the Boltzmann charge distribution [136, 137]. Recent high-resolution DMA measurements of nascent soot in flames shed light on the challenges and potential artifacts affecting studies on soot inception by DMA techniques [13–15]. In the above experiments, an electrometer was used as the particle detector due to the small size of the particles. With the invention of two-stage CPCs, early-stage soot particles with lower concentrations could be detected. By using a 1-nm SMPS system (Model 3936, TSI Inc.), nascent soot particles with sizes around 1.5 nm were observed, which was accompanied by a bimodal particle size distribution when sampled 0.6 cm above the burner [148]. A recent comparison between LII and 1-nm SMPS measurements of nascent soot particles shows good agreement in soot particle size distributions between 2 and 4 nm, demonstrating that LII is also a reliable technique for detecting nascent soot particles below 10 nm [6]. The application of a DMA-APM system for measuring soot particle density and morphology measurement found extremely broad distributions in particle size and mass. These measurements suggest that soot aggregates between 40 and 350 nm, with fractal dimensions ranging from 1.4 to 2.5, are all generated in the same system [125].

By using spectral light absorption measurements, the volume fraction and size distribution of soot and nanoparticles of organic carbon (NOC) were measured [135]. Under sooting conditions, the study observed a good agreement between online (*in situ* and *ex situ*) and offline measurements for sub 10 nm particles, but not so when flames were below the onset of soot. This disagreement requires further studies on the formation mechanisms of NOC and advances in the measurement techniques. The growth mechanism of nascent soot particles was further studied with two-color laser-induced incandescence (2C-LII), with a focus on optical soot properties [9]. The measurements were compared against offline analysis with TEM, and the results showed growing isolated primary soot particles with sizes below 15 nm up to a height of 10 mm above the burner, after which strong soot aggregation occurred and the soot primary particle size remained constant.

AFM has been used to characterize nascent particles produced in sooting flames. Particles with diameters of around 2 nm have been found in the pre-inception region of soot-forming premixed flames, whereas both small nanoparticles and large soot particles have been found in the soot region of the flames. The smaller particles are very flat in shape compared with the bigger ones, possibly due to the different nature of the collected particles [5]. A recent study of nascent soot particles under AFM shows that the collected particles are never spherical [22]. With an increasing flame-equivalent ratio, the particle shape moves from an almost atomically thick object to thicker compounds, indicating a transformation from particles made of small, defective graphene-like sheets to particles containing stacked aromatic layers. The measured attractive and adhesive forces further suggest a continuous increase of the aromatic domains and three-dimensional order within the particles when the flame-equivalent ratio increases.

2.3.1.2 Nanoparticle-synthesizing flame

Flame aerosol reactors are an industrially successful route for synthesizing nanoparticles, providing an alternative for designing and fabricating nanostructures in a one-step and scalable process [83]. The complete oxidation of the synthesis precursor requires a high-temperature environment and removal of residues generated from the gas mixtures. Therefore, flame aerosol reactors typically use a low fuel-to-oxidizer ratio to ensure that the fuel is fully oxidized. Nanoparticles can be produced by premixed flames and turbulent flames, and flame spray pyrolysis is a representative turbulent flame synthesis technique used for synthesizing functional nanoparticles with wide applications [59, 83, 84, 88, 89, 107, 123, 198].

The physical and chemical formation pathways of inorganic oxide nanoparticles during combustion were studied extensively by Wang et al. using *ex situ* methods [166, 171]. In one of these studies, a high-resolution DMA was used to investigate the formation of TiO_2 clusters during flame synthesis. When the synthesis precursor was introduced into a premixed flame, nascent particles with discrete sizes were observed (Figure 2.5a), indicating the formation of stable clusters in early stages [166]. The particle physical size distribution measured by the AFM matched quantitatively well with the mobility size [33]. Coupled with a charged particle remover (CPR), the two-stage CPCs could measure the neutral particle size distributions during combustion, the number fraction of which was significantly lower than that predicted by classical theory [167]. These findings provide new perspectives on the synthesis of functional nanomaterials by changing the ion characteristics in flame environments.

Particle formation during flame synthesis was also investigated using the tandem differential mobility analysis-mass spectrometry (DMA-MS) technique [168]. Measurements in a blank flame detected a large number of sub 2 nm clusters generated from chemical ionization reactions, and these ions played an important role during particle synthesis by attaching to the nascent particles. The simultaneously measured mobility and mass of the flame-generated incipient particles were further compared with the data calculated from empirically determined mass-mobility relationships. The most widely used mass-mobility relationship was presented by Kilpatrick [65] and was further fitted with a function of

$$Z_p = \exp[-0.0347\ln^2(m) - 0.0376\ln(m) + 1.4662] \tag{2.4}$$

where Z_p and m represent the particle electrical mobility (with a unit of $cm^2\ V^{-1}s^{-1}$) and atomic mass (with a unit of Da), respectively. The measurement of incipient particles generated from TiO_2 and SiO_2 synthesis shows that at a same electrical mobility, the actual particle mass is higher than the mass predicted by the empirical relationship. This discrepancy is explained by the fact that the electrical mobility is largely determined by the structure of particles, and particles with similar structures can have different chemical compositions and atomic masses. Due to the existence of relatively heavy species such as silicon and titanium, the flame-generated incipient particles had higher masses. Since researchers often rely on Kilpatrick's relationship to convert the measured mobility to the mass of particles in order to decipher the particle composition [95], this measured result proves that nascent particle mass-mobility relationships will be dependent on the particle chemical composition. Directly using these relationships to convert particle mobility to atomic mass may therefore cause artifacts. The volumetric sizes of the nascent particles are further calculated based on the masses of the particles, which are approximately 0.3 nm smaller than the mobility sizes, agreeing with the findings by [79]. To better predict the mass-mobility relationships, numerical methods were used by researchers to consider the physical collision and potential interaction between the molecular clusters and particles, where a desirable agreement was observed between the calculated and the experimentally measured mass and mobility data [78].

In the field of nanoparticle flame synthesis, an extension of the LII used in soot diagnostics may sound obvious, but it is far from easily accomplished. It has been shown that in many cases the collected radiation signal cannot be plainly assumed to be an incandescence signal [90]. The application of LII in a flame aerosol reactor synthesizing TiO_2 shows that part of the particle synthesis stages can be probed, and that the time decay of the LII signal reflects the particle growth moving away from the precursor injection region [18]. The correlation between mean particle diameter and the time decay along the reaction flame can be separated into different regimes, as expected given the change in gas temperature and particle size distribution. The recent developed PS-LIBS has been shown to be an effective method for probing the early-stage particles [128, 194]. For example, mechanisms involved in flame-assisted-spray-pyrolysis (FASP) synthesis of TiO_2-based functional nanoparticles were studied with the PS-LIBS method [87]. Specifically, the transition from droplets to nanoparticles and the particle growth process in the post-flame region were examined. The emission signal intensity variation revealed the transition from droplets to nanoparticles, indicating that the competition between precursor vaporization and nascent particle formation was the rate determining factor. The PS-LIBS technique has also been used to examine the dynamic behavior of sodium release during pulverized coal combustion, where the gas phase Na release accompanying coal devolatilization was verified *in situ* [183]. The time needed for Na release from particle to gas phases was also derived from the signal.

2.3.2 Furnace aerosol reactors

Furnace aerosol reactors are widely used in the synthesis of functional nanomaterials because the reaction kinetics and downstream particle formation and growth can be facilely controlled by the furnace temperatures [16, 108, 111]. The furnace aerosol reactor can generate sub 10 nm particles by vaporizing a solid material and then rapidly quenching particle growth by introducing clean diluting gas at high flow rates. Measurements of the early-stage particle formation in furnace aerosol reactors showed that particle size distribution evolution (Figure 2.5b) can be used to calculate precursor reaction kinetics [169]. Mass spectroscopic analysis of the early-stage particles proved the difficulty of generating nascent particles with pure compositions [1]. The furnace aerosol reactor could also generate nanoparticles by introducing precursor-containing droplets (also known as spray pyrolysis), where droplet surface reaction and evaporation play significant roles in controlling the properties of the nanoparticles [2, 66]. The DMA-measured size distribution of TiO_2 particles shows that a spray pyrolysis reactor operated at a temperature above 1,000 °C creates faceted, elongated particles and consequently leads to an increased dynamic particle shape factor [2]. *Ex situ* measurement of particle size distribution also assisted the optimization of TeO_2 [187] and gold nanoparticle synthesis [74] in spray pyrolysis reactors. During the synthesis of gold nanoparticles, a clear transformation from droplet-like, spherical gold particle clusters to solid, irregular nanoclusters took place as the temperature increased from 200 to 800 °C. In addition, the primary particle size was below 5 nm in all cases, as shown in TEM images, and no significant sintering took place, even among the particles prepared at high temperatures.

Particles with sizes between 5 and 200 nm generated by furnace aerosol reactors were characterized by LII to determine the primary particle sizes in the case of agglomerated particles. The *in situ* measurement yielded good agreement with DMA and offline analyses [35]. The two-color version of time-resolved laser-induced incandescence (TR-LII) as well as rapid particle probing and transmission electron microscopy (TEM) were applied to measure the size of chain-like iron particles, synthesized by thermally decomposing ironpentacarbonyl ($Fe(CO)_5$) in a hot-wall flow reactor [69]. The study found that for the obtained chain-like agglomerated particle structures, the TR-LII measured size agreed excellently with the TEM determined primary particle size, which is typically between 10 and 50 nm. Other types of nanoparticles prepared by furnace aerosol reactors were examined with AFM and TEM, for example, SnO_2 [39, 64, 119], ZnO [40, 43, 105], and multicomponent particles [80, 134, 138]. These studies investigate the particle formation, growth and film deposition mechanisms during aerosol synthesis.

2.3.3 Other types of aerosol reactors

Here, the working principles of other types of aerosol reactors and corresponding measurements are briefly introduced.

The glowing wire generator, where a material is evaporated from the surface of a heated wire and subsequently quenched by a gas stream, has been commonly applied to produce metallic particles [122]. The generated particles are typically in the range of a few angstroms to a few nanometers, giving them unexpected behaviors [121]. The invention of instruments for measuring early-stage particles made it feasible to accurately measure the size and concentration of these particles, where the mechanisms of metal evaporation and ion attachment were further elucidated. Early-stage particles generated from glowing wire generators (Figure 2.5c) have been used to calibrate the transfer functions and transmission efficiencies of high-resolution DMAs and characterize the detection efficiency of two-stage CPCs under various conditions [60, 61].

A spark discharge generator is an energy-efficient device to produce nanoparticle aerosols in the entire nanometer range (1–100 nm), and beyond [34, 102]. A spark discharge generator generates nanoparticles by charging a shunted capacitor with a high voltage. The subsequently discharged spark evaporates the materials of the electrodes and forms aerosols. Measurements with a high-resolution DMA showed that early-stage silver nanoparticles with discrete sizes (Figure 2.5d) can be produced by spark discharge generators [93]. TEM characterization of the generated particles demonstrated that, due to the high supersaturation of the vapor molecules, the particle formation and growth in a spark discharge generator could be quantified by a versatile model that solely considers the coagulation growth of particles [34].

Laser ablation uses the high energy of a pulsed laser to vaporize a solid precursor that cannot be easily vaporized by a furnace aerosol reactor [98]. Nanoparticles synthesized by laser ablation can also be conveniently characterized by *in situ* techniques, such as LII and LIBS [24, 110, 158, 178]. The structures of the synthesized nanoparticles changed from interconnected chains to dense fibrous aggregates as the number of laser pulse increased. At the same time, the size of the primary particles increased with reactor pressure [116].

2.4 Simulation of particle formation and growth in early stages

Many of the studies referred to above involved significant modeling in studying the early-stage particle formation and growth. These modeling efforts can be categorized based on the dimensions of the modeling systems, for example, quantum mechanics (QM) simulation, molecular dynamics (MD) simulation, and general dynamic simulation, as the simulation length scale increases from below 1 nm to around 1 µm. Based on different assumptions made about the particle growth mechanism, the general dynamic simulations can be further divided into the unimodal model, bimodal model, discrete-sectional model, method of moments, multivariate moment methods, Brownian dynamic model, and Monte Carlo methods.

QM simulation can be used to study the interaction between gas molecules and particle surfaces, which are typically on or below the nanometer scale. This method has been successful in understanding the early stages of soot formation [37, 71, 181, 185].

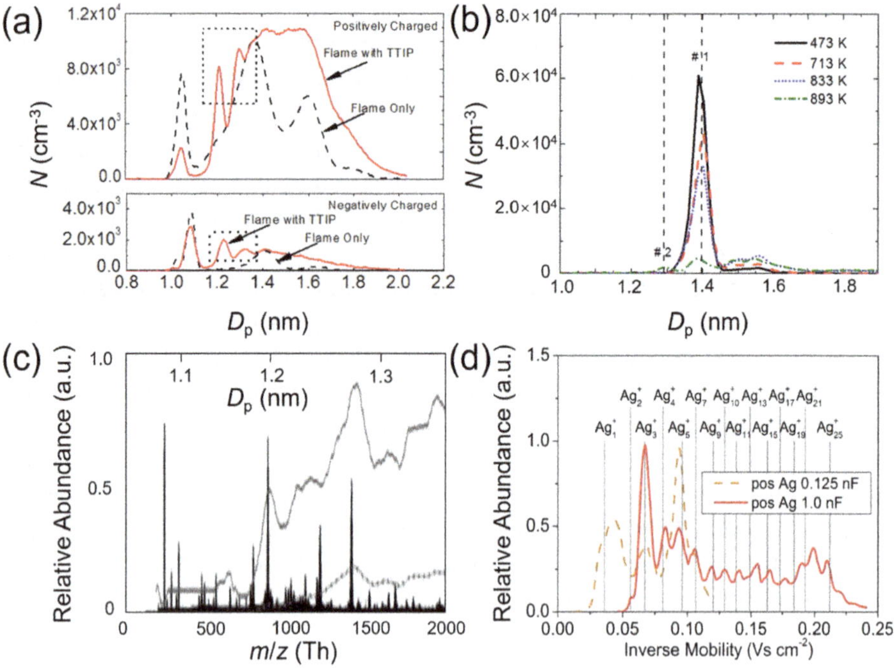

Figure 2.5: Nascent particles measured from a (a) premixed flat-flame aerosol reactor (figure reprinted with permission from Wang et al. [166], Copyright 2014 Elsevier); (b) furnace aerosol reactor (figure reprinted with permission from Wang et al. [169], Copyright 2015 Springer); (c) glowing wire generator (figure reprinted with permission from Kangasluoma et al. (2013); Copyright 2013 Taylor & Francis); and (d) spark discharge generator (figure reprinted with permission from Maisser et al. [93], Copyright 2015 Taylor & Francis).

Simulation results suggest that partitioning of PAHs onto soot depends on the size of the PAH, the planarity of the PAH molecule, and the aromaticity of the compound [72]. By using a combination of QM methods, [181], calculated the reaction kinetics and associated uncertainties of CO and H_2 oxidation at high pressure and temperatures. Moreover, the QM simulation can predict the enthalpy of formation for metal oxides formed during combustion, which is substantially different from the equilibrium composition from thermochemistry available in the literature [147].

MD simulations have unraveled the evolution of the particle morphology of various nanomaterials during particle coagulation [176, 192], sintering [10, 27, 47, 70, 184], and interaction with surrounding ions and molecules [42, 76, 78]. The simulations typically cover a size scale below 5 nm, limited by computational power. For early-stage particles, the distributions of atoms are asymmetric at the surface of the particle, resulting in permanent dipoles that are approximately proportional to the surface area in magnitude [176]. MD simulation shows that the interaction among dipoles in each particle significantly enhances coagulation, and a temperature increase weakens the time-averaged dipole moment due to thermal fluctuations [192]. MD simulation also quantified and elucidated the growth of TiO_2 nanoparticles by sintering (Figure 2.6), which is a critical step during their large-scale manufacture and processing. The characteristic sintering time of particle pairs can be determined by tracing their surface area evolution during coalescence, and particle crystallinity can be quantified by the system's degree of disorder [47]. The atoms at the grain boundaries are found to be amorphous, especially during particle adhesion and sintering, when grains of different orientation are formed. Simulation results show that mobile ions from the particle surface fill in the initially concave space between nanoparticles (surface diffusion) forming the final, fully coalesced, spherical particle with minimal displacement of inner ions (grain boundary diffusion) [10]. Building on the MD simulation of a particle under equilibrium conditions, the structure of the particle can be used to simulate its interaction with surrounding ions and gas molecules [76, 77]. The collision cross section derived from the simulation is further related to the mobility of the particles, which can be measured with high-resolution DMAs. The good agreement between the calculated and measured mobility of particles below 3 nm demonstrated the capability of such simulation approaches in predicting particle physical properties [94].

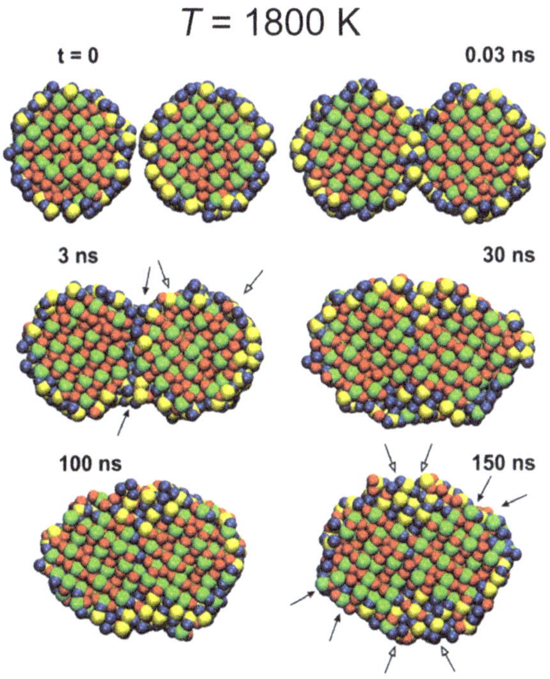

Figure 2.6: MD simulation of the sintering processes for TiO_2 dioxide nanoparticles. The initial size of the simulated particles is 3 nm, and the color represents Ti (green), O (red), and surface ions (yellow and blue). (Figure reprinted with permission from Buesser et al. [10], Copyright 2011 ACS.).

There are a plethora of modeling methods for predicting particle formation and growth using the general dynamic equations. General dynamic simulations can cover a wide range of particle sizes, from early-stage particles to submicron particles, because detailed information on particle morphology and chemical composition is neglected or assumed to be uniform. The general dynamic equation of particle size distribution function (n) has the form of

$$\frac{\partial n}{\partial t}+\nabla\cdot n\mathbf{v}+\frac{\partial I}{\partial t}=\nabla\cdot D\nabla n+\frac{1}{2}\int\beta(\tilde{v},v-\tilde{v})n(\tilde{v})n(v-\tilde{v})d\tilde{v}-\int\beta(v,\tilde{v})n(v)n(\tilde{v})d\tilde{v}-\nabla\cdot n\mathbf{c} \tag{2.5}$$

In the equation, $\mathbf{v}$ is the flow velocity, I is the particle source term, D is the diffusion coefficient, β is the coagulation coefficient between two particles, and $\mathbf{c}$ is the particle velocity.

Because such equations are nonlinear, assumptions are made to simplify the calculation process. For example, a unimodal model assumes that particles exist in a single size mode, and the particle growth by condensation and coagulation is lumped and solved by mass conservation [36]. In a bimodal model, two discrete modes are introduced to represent a bimodal size distribution: a monodisperse mode accounts for the formation of new particles in nucleation, and an accumulation mode accounts for the growth of larger particles [56].

The general dynamic equations, on the other hand, can be solved in discrete form, accounting separately for each cluster or particle size in terms of the number of molecules the particle contains, beginning with monomer, dimer, trimer, etc. Alternatively, they can be solved in continuous form, wherein the distribution of particles with respect to size is represented by a continuous function. The former approach provides an accurate description of the evolution of small clusters, for example, the initial formation of new particles by homogeneous nucleation. The observation of particles with discrete size distributions below 3 nm indicates that the discrete method may be an appropriate method for describing nascent particle formation. However, as particles further grow by condensation or coagulation with smaller particles, the addition of volume becomes negligible, and the total number of cluster sizes that would be needed to describe aerosols extending to measurable sizes in the discrete form becomes immense. In this situation, the latter approach is natural for describing aerosols that include a broad range of particle sizes. For example, the sectional modeling of particle formation and growth during the combustion synthesis of TiO_2 particles yielded good agreement with experimental measurements [153]. A combination of the two approaches, called discrete-sectional modeling, could further enhance the accuracy of the models, and capture the evolution of particle dynamics during early-stage particle formation [75, 174].

The method of moments is a class of techniques for tracking preselected moments of a particle size distribution directly in space and time without explicit knowledge of the particle size distribution itself. The particle size distributions can be assumed to be lognormal, and the calculation of the particle evolution could be greatly simplified by the moments of the size distribution. One advantage of the method is its great computational efficiency, because only a few variables need to be integrated to accurately obtain important integral properties [4, 182]. However, simulation by the method of moments provides particle size information only. Therefore, the major application for multivariate moment method with multiple state variables is modeling the dynamics of generally mixed, multicomponent particle populations as well as particle populations having both mixed composition and complex shapes [101, 173, 179, 180].

Brownian dynamic simulation tracks the motion of particles as a function of time and can be used to predict the growth of early-stage particles, which cannot be easily described by continuum models. The simulation solves the stochastic Langevin equation, incorporating the random Brownian motion of colloidal particles. The Langevin equation describing the particle movement can be simplified as

$$r_{\mathrm{p},t+\Delta t} = r_{\mathrm{p},t} + \left(\frac{\Delta t}{f}\right) F_{\mathrm{ext}} + B_3 \tag{2.6}$$

In this equation, $r_{\mathrm{p},t}$ is the location of the particles at time t, Δt is any time interval, f is the friction factor calculated by $f = 3\pi\mu D_p / C$, F_{ext} is the external force, and B_3 is a Gaussian random variable. B_3 satisfies that $B_3 = 0$ and $B_3^2 = 6\frac{kT}{f}\Delta t$. Such Brownian dynamic modeling methods have been used to study ion-particle interaction [44, 45], particle morphology during coagulation [73], and particle deposition [156, 163]. Monte Carlo simulation can be built upon Brownian dynamic simulation by introducing a large number of particles in the model, so that the physical properties of the particles can be obtained with high reproducibility [26, 44]. This type of Monte Carlo simulation is called "time-driven Monte Carlo" by 92. On the other hand, if the probability of an event, for example, coagulation, condensation, or sintering, is known, one can use "event-driven Monte Carlo" to study the aerosol dynamics [91, 92, 197].

Particle-ion interactions in aerosol reactors can exert a significant influence on nascent particle formation and growth. Experiments found that external electrical fields significantly influence the morphology of the synthesized particles [160]. However, most of the introduced models neglect particle charging or assume an equilibrium or steady-state charge distribution, whereas aerosol reactors form particles in a non-equilibrium and transient manner. A few studies demonstrate that simultaneous particle charging and coagulation may be different from the particle dynamics predicted by commonly used models. In these models, the ion-particle combination coefficients ($\eta_i^{\pm}$) are introduced to the particle coagulation system. The ion-particle combination coefficient $\eta_i^{\pm}$ of positive (+) and negative (–) ions with a particle carrying i elementary units of charge [127] can be derived as

$$\eta_i^{\pm} = \frac{\beta_i^{\pm} \exp[-\varphi_i(\delta^{\pm})/kT]}{1 + \exp[-\varphi_i(\delta^{\pm})/kT]\left[\beta_i^{\pm}/4\pi D^{\pm}\right] \int_0^1 \exp[-\varphi_i(\delta^{\pm})/kT] dx} \tag{2.7}$$

In this equation, $\beta_i^{\pm}$ is the proportionality between the ion flux onto a particle and the ion concentration, φ_i is the electric potential surrounding the particle, $\delta^{\pm}$ is the radius of the limiting sphere (inside which the movement of ions is guided by the electrostatic potential between the ion and the particle), k is the Boltzmann's constant, T is temperature, and $D^{\pm}$ is the diffusion coefficient of the ions. $\beta_i^{\pm}$ and $D^{\pm}$ are functions of ion mass and mobility respectively. In the presence of electrostatic interaction between charged particles, the coagulation coefficient between particles is given by

$$\beta_{\text{coag}}^{i,j}\left(D_{\text{pi}}, D_{\text{pj}}\right) = \frac{\beta_{\text{Brownian}}\left(D_{\text{pi}}, D_{\text{pj}}\right)}{W\left(z_i, z_j\right)} \tag{2.8}$$

where $\beta_{\text{Brownian}}\left(D_{\text{pi}}, D_{\text{pj}}\right)$ is the coagulation coefficient due to Brownian motion without charge involvement, D_{pi} and D_{pj} are diameters of coagulating particles, z_i and z_j are the number of charges on the particles, and W is a correction factor as a function of z_i and z_j. $\beta_{\text{Brownian}}\left(D_{\text{pi}}, D_{\text{pj}}\right)$ is further calculated by

$$\beta_{brownian}\left(D_{\text{pi}}, D_{\text{pj}}\right) = 2\pi\left(D_{\text{pi}} + D_{\text{pj}}\right)\left(D_i + D_j\right)\left\{\frac{D_{\text{pi}} + D_{\text{pj}}}{D_{\text{pi}} + D_{\text{pj}} + 2(g_i^2 + g_j^2)^{1/2}} + \frac{8\left(D_i + D_j\right)}{\left(D_{\text{pi}} + D_{\text{pj}}\right)\left(c_i^2 + c_j^2\right)^{1/2}}\right\}^{-1} \tag{2.9}$$

where c_i is the mean particle velocity, $c_i = \left(8kT/(\pi m_i)\right)^{1/2}$. m_i is the mass of the particle, D_i is the diffusion coefficient of the particle, where $D_i = \frac{kT}{3\pi\mu D_{\text{pi}}}\left\{\frac{5 + 4Kn_i + 6Kn_i^2 + 18Kn_i^3}{5 - Kn_i + (8+\pi)Kn_i^2}\right\}$, and Kn_i is the particle Knudsen number. Furthermore, $g_i = \frac{1}{3D_{\text{pi}}l_i}\left\{\left(D_{\text{pi}} + l_i\right)^3 - \left(D_{\text{pi}}^2 + l_i^2\right)^{3/2}\right\} - D_{\text{pi}}$ and $l_i = 8D_i/(\pi c_i)$. The population balance of ions and particles with different charge states then can be constructed, considering the ion-particle and particle-particle interactions.

Fujimoto et al. [38] calculated the unipolar ion charging and particle coagulation during aerosol formation from titanium isopropoxide thermal decomposition at temperatures below 500 °C, and predicted that the particles' average size, standard deviation, and concentration were affected greatly. Sectional and method of moments modeling show that bipolar charging enhances particle coagulation under 700 °C [120]. The simulation of particle charging and coagulation of radioactive and non-radioactive particles in the atmosphere found that the mutual effects of the two mechanisms indeed alter the particle size distribution [67]. The interactions among ions and particles under high temperatures were also modeled via the method of moments by simplifying the ion-particle combination coefficient to a polynomial expression as a function of particle size [139]. The results compare well against a monodisperse model that simulates simultaneous particle charging and coagulation. These modeling results have shown that the influence of charging on particle growth dynamics was more prominent when the ion concentration was comparable to or higher than the particle concentrations, a condition that may be encountered in flame synthesis and solid fuel-burning [171].

2.5 Summary

The properties of nanoparticles generated in aerosol reactors depend in a complex way on the reaction kinetics, heat transfer, and mass transfer. In order to understand the early-stage particle formation mechanisms, analysis of the final product alone by offline techniques cannot fully clarify their formation pathways, for such purposes such as modifying material properties and controlling pollutant formation. Online measurements, both *in situ* and *ex situ*, can provide a wide variety of information that can be temporally and spatially resolved to understand the initial stages of particle formation. This chapter introduced several characterization methods and their findings for early-stage particles. As can be seen, many suitable techniques are well established in combustion science and atmospheric science, and they can be transferred to aerosol reactors with some adjustment.

Developing aerosol dynamic models greatly assists our understanding of particle formation pathways in materials synthesis and pollution control. To accurately simulate early-stage particles, more efforts are needed in coupling the detailed chemical reaction kinetics, multidimensional particle size distributions, and computational fluid dynamics simulations. The application and validation of the models against experimental observations under various reaction conditions will provide useful information about particle formation in early stages.

References

[1] Ahonen, L.R., Kangasluoma, J., Lammi, J., Lehtipalo, K., Hämeri, K., Petäjä, T., Kulmala, M. (2017). First measurements of the number size distribution of 1–2 nm aerosol particles released from manufacturing processes in a cleanroom environment. *Aerosol Science and Technology* 51(6), 685–693.

[2] Ahonen, P.P., Joutsensaari, J., Richard, O., Tapper, U., Brown, D.P., Jokiniemi, J.K., Kauppinen, E.I. (2001). Mobility size development and the crystallization path during aerosol decomposition synthesis of TiO2 particles. *Journal of Aerosol Science* 32:615–630.

[3] Akimoto, H. (2003). Global air quality and pollution. Science 302:1716–1719.

[4] Bae, S.Y., Jung, C.H., Kim, Y.P. (2010). Derivation and verification of an aerosol dynamics expression for the below-cloud scavenging process using the moment method. *Journal of Aerosol Science* 41:266–280.

[5] Barone, A., d'Alessio, A., d'Anna, A. (2003). Morphological characterization of the early process of soot formation by atomic force microscopy. *Combustion and Flame* 132:181–187.

[6] Betrancourt, C., Liu, F., Desgroux, P., Mercier, X., Faccinetto, A., Salamanca, M., Ruwe, L., Kohse-Höinghaus, K., Emmrich, D., Beyer, A. (2017). Investigation of the size of the incandescent incipient soot particles in premixed sooting and nucleation flames of n-butane using LII, HIM, and 1 nm-SMPS. *Aerosol Science and Technology* 51:916–935.

[7] Biswas, P., Thimsen, E. (2011). High temperature aerosols: Measurement and deposition of nanoparticle films in Aerosol measurement: principles, techniques, and applications. Kulkarni, Pramod, Paul A. Baron, and Klaus Willeke, eds. John Wiley & Sons, 2011. 723–738.

[8] Biswas, P., Wang, Y., Attoui, M. (2018). Sub-2nm particle measurement in high-temperature aerosol reactors: A review. *Current Opinion in Chemical Engineering* 21:60–66.

[9] Bladh, H., Johnsson, J., Olofsson, N.-E., Bohlin, A., Bengtsson, P.-E.: Optical soot characterization using two-color laser-induced incandescence (2C-LII) in the soot growth region of a premixed flat flame. *Proceedings of the Combustion Institute* 33:641–648, 2011.

[10] Buesser, B., Grohn, A., Pratsinis, S.E. (2011). Sintering rate and mechanism of TiO2 nanoparticles by molecular dynamics. *The Journal of Physical Chemistry C* 115:11030–11035.

[11] Camacho, J., Lieb, S., Wang, H. (2013). Evolution of size distribution of nascent soot in n-and i-butanol flames. *Proceedings of the Combustion Institute* 34:1853–1860.

[12] Camacho, J., Liu, C., Gu, C., Lin, H., Huang, Z., Tang, Q., You, X., Saggese, C., Li, Y., Jung, H. (2015). Mobility size and mass of nascent soot particles in a benchmark premixed ethylene flame. *Combustion and Flame* 162:3810–3822.

[13] Carbone, F., Attoui, M., Gomez, A. (2016). Challenges of measuring nascent soot in flames as evidenced by high-resolution differential mobility analysis. *Aerosol Science and Technology* 50:740–757.

[14] Carbone, F., Moslih, S., Gomez, A. (2017a). Probing gas-to-particle transition in a moderately sooting atmospheric pressure ethylene/air laminar premixed flame. Part I: Gas phase and soot ensemble characterization. *Combustion and Flame*, 181, 315–328.

[15] Carbone, F., Moslih, S., Gomez, A. (2017b). Probing gas-to-particle transition in a moderately sooting atmospheric pressure ethylene/air laminar premixed flame. Part II: Molecular clusters and nascent soot particle size distributions. *Combustion and Flame*, 181, 329–341.

[16] Cho, K., Biswas, P. (2006). Sintering rates for pristine and doped titanium dioxide determined using a tandem differential mobility analyzer system. *Aerosol Science and Technology* 40:309–319.

[17] Choi, K., Kim, J., Ko, A., Myung, C.-L., Park, S., Lee, J. (2013). Size-resolved engine exhaust aerosol characteristics in a metal foam particulate filter for GDI light-duty vehicle. *Journal of Aerosol Science* 57:1–13.

[18] Cignoli, F., Bellomunno, C., Maffi, S., Zizak, G. (2009). Laser-induced incandescence of titania nanoparticles synthesized in a flame. *Applied Physics B* 96:593–599.

[19] D'alessio, A., D'Anna, A., Minutolo, P., Sgro, L., Violi, A. (2000). On the relevance of surface growth in soot formation in premixed flames. *Proceedings of the Combustion Institute* 28:2547–2554.

[20] D'Anna, A. (2009). Combustion-formed nanoparticles. *Proceedings of the Combustion Institute* 32:593–613.

[21] Daun, K., Sipkens, T., Titantah, J., Karttunen, M. (2013). Thermal accommodation coefficients for laser-induced incandescence sizing of metal nanoparticles in monatomic gases. *Applied Physics B* 112:409–420.

[22] De Falco, G., Commodo, M., Minutolo, P., D'Anna, A. (2015). Flame-formed carbon nanoparticles: Morphology, interaction forces, and hamaker constant from AFM. *Aerosol Science and Technology* 49:281–289.

[23] De Filippo, A., Sgro, L., Lanzuolo, G., D'Alessio, A. (2009). Probe measurements and numerical model predictions of evolving size distributions in premixed flames. *Combustion and Flame* 156:1744–1754.
[24] De Giacomo, A., Dell'Aglio, M., Gaudiuso, R., Koral, C., Valenza, G. (2016). Perspective on the use of nanoparticles to improve LIBS analytical performance: Nanoparticle enhanced laser induced breakdown spectroscopy (NELIBS). *Journal of Analytical Atomic Spectrometry* 31:1566–1573.
[25] de La Mora, J.F., Kozlowski, J. (2013). Hand-held differential mobility analyzers of high resolution for 1–30 nm particles: Design and fabrication considerations. *Journal of Aerosol Science* 57:45–53.
[26] Debry, E., Sportisse, B., Jourdain, B. (2003). A stochastic approach for the numerical simulation of the general dynamics equation for aerosols. *Journal of Computational Physics* 184:649–669.
[27] Ding, L., Davidchack, R.L., Pan, J. (2009). A molecular dynamics study of sintering between nanoparticles. *Computational Materials Science* 45:247–256.
[28] Dockery, D.W. (2001). Epidemiologic evidence of cardiovascular effects of particulate air pollution. *Environmental Health Perspectives* 109:483–486.
[29] Dreier, T., Schulz, C. (2016). Laser-based diagnostics in the gas-phase synthesis of inorganic nanoparticles. *Powder Technology* 287:226–238.
[30] Dunne, E.M., Gordon, H., Kürten, A., Almeida, J., Duplissy, J., Williamson, C., Ortega, I.K., Pringle, K.J., Adamov, A., Baltensperger, U. (2016). Global atmospheric particle formation from CERN CLOUD measurements. *Science* 354:1119–1124.
[31] Ehara, K.: Aerosol mass spectrometer and method of classifying aerosol particles according to specific mass, Google Patents, 1995.
[32] Ehara, K., Hagwood, C., Coakley, K.J. (1996). Novel method to classify aerosol particles according to their mass-to-charge ratio – Aerosol particle mass analyser. *Journal of Aerosol Science* 27:217–234.
[33] Fang, J., Wang, Y., Attoui, M., Chadha, T.S., Ray, J.R., Wang, W.-N., Jun, Y.-S., Biswas, P. (2014). Measurement of sub-2 nm clusters of pristine and composite metal oxides during nanomaterial synthesis in flame aerosol reactors. *Analytical Chemistry* 86:7523–7529.
[34] Feng, J., Huang, L., Ludvigsson, L., Messing, M.E., Maisser, A., Biskos, G., Schmidt-Ott, A. (2015). General approach to the evolution of singlet nanoparticles from a rapidly quenched point source. *The Journal of Physical Chemistry C* 120:621–630.
[35] Filippov, A., Markus, M., Roth, P. (1999). In-situ characterization of ultrafine particles by laser-induced incandescence: Sizing and particle structure determination. *Journal of Aerosol Science* 30:71–87.
[36] Friedlander, S.K. (1977). Smoke, Dust and Haze: Fundamentals of Aerosol Behavior, Wiley-Interscience, New York, 333, 1977.
[37] Froudakis, G.E. (2001). Hydrogen interaction with single-walled carbon nanotubes: A combined quantum-mechanics/molecular-mechanics study. *Nano Letters* 1:179–182.
[38] Fujimoto, T., Kuga, Y., Pratsinis, S.E., Okuyama, K. (2003). Unipolar ion charging and coagulation during aerosol formation by chemical reaction. *Powder Technology* 135:321–335.
[39] Gaiduk, P., Kozjevko, A., Prokopjev, S., Tsamis, C., Larsen, A.N. (2008). Structural and sensing properties of nanocrystalline SnO 2 films deposited by spray pyrolysis from a SnCl 2 precursor. *Applied Physics A* 91:667–670.
[40] Gaikwad, R.S., Bhande, S.S., Mane, R.S., Pawar, B.N., Gaikwad, S.L., Han, S.-H., Joo, O.-S. (2012). Roughness-based monitoring of transparency and conductivity in boron-doped ZnO thin films prepared by spray pyrolysis. *Materials Research Bulletin* 47:4257–4262.

[41] Gao, Q., Li, S., Yuan, Y., Zhang, Y., Yao, Q. (2015). Ultrafine particulate matter formation in the early stage of pulverized coal combustion of high-sodium lignite. *Fuel* 158:224–231.

[42] Gao, Y., Liu, J., Shen, J., Wu, Y., Zhang, L. (2014). Influence of various nanoparticle shapes on the interfacial chain mobility: A molecular dynamics simulation. *Physical Chemistry Chemical Physics* 16:21372–21382.

[43] Godbole, B., Badera, N., Shrivastava, S., Jain, D., Ganesan, V. (2011). Growth mechanism of ZnO films deposited by spray pyrolysis technique. *Materials Sciences and Applications* 2:643.

[44] Gopalakrishnan, R., Meredith, M.J., Larriba-Andaluz, C., Hogan Jr, C.J. (2013a). Brownian dynamics determination of the bipolar steady state charge distribution on spheres and non-spheres in the transition regime. *Journal of Aerosol Science* 63:126–145.

[45] Gopalakrishnan, R., Thajudeen, T., Ouyang, H., Hogan Jr, C.J. (2013b). The unipolar diffusion charging of arbitrary shaped aerosol particles. *Journal of Aerosol Science* 64:60–80.

[46] Goudeli, E., Gröhn, A.J., Pratsinis, S.E. (2016). Sampling and dilution of nanoparticles at high temperature. *Aerosol Science and Technology* 50:591–604.

[47] Goudeli, E., Pratsinis, S.E. (2016). Crystallinity dynamics of gold nanoparticles during sintering or coalescence. *AIChE Journal* 62:589–598.

[48] Gröhn, A.J., Pratsinis, S.E., Wegner, K. (2012). Fluid-particle dynamics during combustion spray aerosol synthesis of ZrO2. *Chemical Engineering Journal* 191:491–502.

[49] Harrison, R.M., Hester, R.E., Querol, X. (2016). Airborne Particulate Matter: Sources, Atmospheric Processes and Health, Royal Society of Chemistry.

[50] Hering, S.V., Lewis, G.S., Spielman, S.R., Eiguren-Fernandez, A., Kreisberg, N.M., Kuang, C., Attoui, M. (2017). Detection near 1-nm with a laminar-flow, water-based condensation particle counter. *Aerosol Science and Technology* 51:354–362.

[51] Herrmann, W., Eichler, T., Bernardo, N., Fernández de la Mora, J. Turbulent transition arises at reynolds number 35,000 in a short vienna type DMA with a large laminarization inlet, in Abstract AAAR Conference, 15B5, 2000.

[52] Hildemann, L., Cass, G., Markowski, G. (1989). A dilution stack sampler for collection of organic aerosol emissions: Design, characterization and field tests. *Aerosol Science and Technology* 10:193–204.

[53] Hu, Y., Jiang, H., Li, Y., Wang, B., Zhang, L., Li, C., Wang, Y., Cohen, T., Jiang, Y., Biswas, P. (2017). Engineering the outermost layers of TiO2 nanoparticles using in situ Mg doping in a flame aerosol reactor. *AIChE Journal* 63:870–880.

[54] Huffman, J., Sinha, B., Garland, R., Snee-Pollmann, A., Gunthe, S., Artaxo, P., Martin, S., Andreae, M., Pöschl, U. (2012). Size distributions and temporal variations of biological aerosol particles in the Amazon rainforest characterized by microscopy and real-time UV-APS fluorescence techniques during AMAZE-08. *Atmospheric Chemistry and Physics* 12:11997–12019.

[55] Iida, K., Stolzenburg, M.R., McMurry, P.H. (2009). Effect of working fluid on sub-2 nm particle detection with a laminar flow ultrafine condensation particle counter. *Aerosol Science and Technology* 43:81–96.

[56] Jeong, J.I., Choi, M. (2004). A bimodal moment model for the simulation of particle growth. *Journal of Aerosol Science* 35:1071–1090.

[57] Jiang, J., Attoui, M., Heim, M., Brunelli, N.A., McMurry, P.H., Kasper, G., Flagan, R.C., Giapis, K., Mouret, G. (2011). Transfer functions and penetrations of five differential mobility analyzers for sub-2 nm particle classification. *Aerosol Science and Technology* 45:480–492.

[58] Junninen, H., Ehn, M., Petäjä, T., Luosujärvi, L., Kotiaho, T., Kostiainen, R., Rohner, U., Gonin, M., Fuhrer, K., Kulmala, M. (2010). A high-resolution mass spectrometer to measure atmospheric ion composition. *Atmospheric Measurement Techniques* 3:1039–1053.

[59] Kammler, H.K., Mädler, L., Pratsinis, S.E. (2001). Flame synthesis of nanoparticles. *Chemical Engineering & Technology: Industrial Chemistry-Plant Equipment-Process Engineering-Biotechnology* 24:583–596.

[60] Kangasluoma, J., Attoui, M., Junninen, H., Lehtipalo, K., Samodurov, A., Korhonen, F., Sarnela, N., Schmidt-Ott, A., Worsnop, D., Kulmala, M. (2015). Sizing of neutral sub 3nm tungsten oxide clusters using Airmodus Particle Size Magnifier. *Journal of Aerosol Science* 87:53–62.

[61] Kangasluoma, J., Junninen, H., Lehtipalo, K., Mikkilä, J., Vanhanen, J., Attoui, M., Sipilä, M., Worsnop, D., Kulmala, M., Petäjä, T. (2013). Remarks on ion generation for CPC detection efficiency studies in sub-3-nm size range. *Aerosol Science and Technology* 47:556–563.

[62] Kangasluoma, J., Kuang, C., Wimmer, D., Rissanen, M., Lehtipalo, K., Ehn, M., Worsnop, D., Wang, J., Kulmala, M., Petäjä, T. (2014). Sub-3 nm particle size and composition dependent response of a nano-CPC battery. *Atmospheric Measurement Techniques* 7:689–700.

[63] Kasper, M., Siegmann, K., Sattler, K. (1997). Evaluation of an in situ sampling probe for its accuracy in determining particle size distributions from flames. *Journal of Aerosol Science* 28:1569–1578.

[64] Kaur, J., Kumar, R., Bhatnagar, M. (2007). Effect of indium-doped SnO2 nanoparticles on NO2 gas sensing properties. *Sensors and Actuators. B, Chemical* 126:478–484.

[65] Kilpatrick, W.: An experimental mass-mobility relation for ions in air at atmospheric pressure, in Proc. Annu. Conf. Massspectrosc, 320–325, 1971.

[66] Kim, J.H., Germer, T.A., Mulholland, G.W., Ehrman, S.H. (2002). Size-monodisperse metal nanoparticles via hydrogen-free spray pyrolysis. *Advanced Materials* 14:518–521.

[67] Kim, Y.-H., Yiacoumi, S., Nenes, A., Tsouris, C. (2016). Charging and coagulation of radioactive and nonradioactive particles in the atmosphere. *Atmospheric Chemistry and Physics* 16:3449–3462.

[68] Knutson, E., Whitby, K. (1975). Aerosol classification by electric mobility: Apparatus, theory, and applications. *Journal of Aerosol Science* 6:443–451.

[69] Kock, B.F., Kayan, C., Knipping, J., Orthner, H.R., Roth, P.: Comparison of LII and TEM sizing during synthesis of iron particle chains. Proceedings of the Combustion Institute 30:1689–1697, 2005.

[70] Koparde, V.N., Cummings, P.T. (2005). Molecular dynamics simulation of titanium dioxide nanoparticle sintering. *The Journal of Physical Chemistry. B* 109:24280–24287.

[71] Kubicki, J. (2000). Molecular mechanics and quantum mechanical modeling of hexane soot structure and interactions with pyrene. *Geochemical Transactions* 1:41.

[72] Kubicki, J.D. (2006). Molecular simulations of benzene and PAH interactions with soot. *Environmental Science & Technology* 40:2298–2303.

[73] Kulkarni, P., Biswas, P. (2004). A Brownian dynamics simulation to predict morphology of nanoparticle deposits in the presence of interparticle interactions. *Aerosol Science and Technology* 38:541–554.

[74] Lähde, A., Koshevoy, I., Karhunen, T., Torvela, T., Pakkanen, T.A., Jokiniemi, J. (2014). Aerosol-assisted synthesis of gold nanoparticles. *Journal of Nanoparticle Research* 16:2716.

[75] Landgrebe, J.D., Pratsinis, S.E. (1990). A discrete-sectional model for particulate production by gas-phase chemical reaction and aerosol coagulation in the free-molecular regime. *Journal of Colloid and Interface Science* 139:63–86.

[76] Larriba-Andaluz, C., Hogan, C.J. Jr (2014). Collision cross section calculations for polyatomic ions considering rotating diatomic/linear gas molecules. *The Journal of Chemical Physics* 141:194107.

[77] Larriba, C., Hogan, C.J. Jr (2013a). Free molecular collision cross section calculation methods for nanoparticles and complex ions with energy accommodation. *Journal of Computational Physics* 251:344–363.

[78] Larriba, C., Hogan, C.J. Jr (2013b). Ion mobilities in diatomic gases: Measurement versus prediction with non-specular scattering models. *The Journal of Physical Chemistry. A* 117:3887–3901.
[79] Larriba, C., Hogan, C.J. Jr, Attoui, M., Borrajo, R., Garcia, J.F., de La Mora, J.F. (2011). The mobility–volume relationship below 3.0 nm examined by tandem mobility–mass measurement. *Aerosol Science and Technology* 45:453–467.
[80] Lee, C.-S., Hsu, W.-C., Chou, B.-Y., Liu, H.-Y., Yang, C.-L., Sun, W.-C., Wei, S.-Y., Yu, S.-M., Wu, C.-L. (2015). Investigations of TiO 2–AlGaN/GaN/Si-passivated HFETs and MOS-HFETs using ultrasonic spray pyrolysis deposition. *IEEE Transactions on Electron Devices* 62:1460–1466.
[81] Lehtipalo, K., Leppä, J., Kontkanen, J., Kangasluoma, J., Franchin, A., Wimmer, D., Schobesberger, S., Junninen, H., Petäjä, T., Sipilä, M.: Methods for determining particle size distribution and growth rates between 1 and 3 nm using the Particle Size Magnifier, 2014.
[82] Li, J., Leavey, A., Wang, Y., O'Neil, C., Wallace, M.A., Burnham, C.-A.D., Boon, A.C., Babcock, H., Biswas, P. (2018a). Comparing the performance of 3 bioaerosol samplers for influenza virus. *Journal of Aerosol Science* 115:133–145.
[83] Li, S., Ren, Y., Biswas, P., Stephen, D.T. (2016). Flame aerosol synthesis of nanostructured materials and functional devices: Processing, modeling, and diagnostics. *Progress in Energy and Combustion Science* 55:1–59.
[84] Li, W., Hu, Y., Jiang, H., Jiang, Y., Wang, Y., Huang, S., Biswas, P., Li, C. (2018b). Fluxing template-assisted synthesis of sponge-like Fe2O3 microspheres toward efficient catalysis for CO oxidation. *Applied Surface Science* 444:763–771.
[85] Li, Z., Wang, Y., Lu, Y., Biswas, P. (2019). Investigation of aerosol and gas emissions from a coal-fired power plant under various operating conditions. *Journal of the Air & Waste Management Association* 69:34–46.
[86] Liu, B., Pui, D., Rubow, K., Szymanski, W. (1985). Electrostatic effects in aerosol sampling and filtration. *The Annals of Occupational Hygiene* 29:251–269.
[87] Liu, C., Li, S., Zong, Y., Yao, Q., Stephen, D.T.: Laser-based investigation of the transition from droplets to nanoparticles in flame-assisted spray synthesis of functional nanoparticles. Proceedings of the Combustion Institute 36:1109–1117, 2017.
[88] Lu, L., Hu, Y., Jiang, H., Wang, Y., Jiang, Y., Huang, S., Niu, X., Biswas, P., Li, C. (2018). Multi-shelled LiMn 1.95 Co 0.05 O 4 cages with a tunable Mn oxidation state for ultra-high lithium storage. *New Journal of Chemistry* 42:3953–3960.
[89] Mädler, L., Kammler, H., Mueller, R., Pratsinis, S.E. (2002). Controlled synthesis of nanostructured particles by flame spray pyrolysis. *Journal of Aerosol Science* 33:369–389.
[90] Maffi, S., Cignoli, F., Bellomunno, C., De Iuliis, S., Zizak, G. (2008). Spectral effects in laser induced incandescence application to flame-made titania nanoparticles. *Spectrochimica Acta. Part B: Atomic Spectroscopy* 63:202–209.
[91] Maisels, A., Jordan, F., Kruis, F.E., Fissan, H. (2003). A study of nanoparticle aerosol charging by Monte Carlo simulations. *Journal of Nanoparticle Research* 5:225–235.
[92] Maisels, A., Kruis, F.E., Fissan, H. (2004). Direct simulation Monte Carlo for simultaneous nucleation, coagulation, and surface growth in dispersed systems. *Chemical Engineering Science* 59:2231–2239.
[93] Maisser, A., Barmpounis, K., Attoui, M., Biskos, G., Schmidt-Ott, A. (2015). Atomic cluster generation with an atmospheric pressure spark discharge generator. *Aerosol Science and Technology* 49:886–894.
[94] Maißer, A., Thomas, J.M., Larriba-Andaluz, C., He, S., Hogan, C.J. Jr (2015). The mass–mobility distributions of ions produced by a Po-210 source in air. *Journal of Aerosol Science* 90:36–50.

[95] Mäkelä, J.M., Jokinen, V., Mattila, T., Ukkonen, A., Keskinen, J. (1996). Mobility distribution of acetone cluster ions. *Journal of Aerosol Science* 27:175–190.
[96] Maricq, M. (2004). Size and charge of soot particles in rich premixed ethylene flames. *Combustion and Flame* 137:340–350.
[97] Maricq, M.M., Harris, S.J., Szente, J.J. (2003). Soot size distributions in rich premixed ethylene flames. *Combustion and Flame* 132:328–342.
[98] Marine, W., Patrone, L., Luk'yanchuk, B., Sentis, M. (2000). Strategy of nanocluster and nanostructure synthesis by conventional pulsed laser ablation. *Applied Surface Science* 154:345–352.
[99] Markowski, G.R. (1987). Improving Twomey's algorithm for inversion of aerosol measurement data. *Aerosol Science and Technology* 7:127–141.
[100] Mauderly, J.L., Jones, R.K., Griffith, W.C., Henderson, R.F., McClellan, R.O. (1987). Diesel exhaust is a pulmonary carcinogen in rats exposed chronically by inhalation. *Fundamental and Applied Toxicology* 9:208–221.
[101] McGraw, R. (1997). Description of aerosol dynamics by the quadrature method of moments. *Aerosol Science and Technology* 27:255–265.
[102] Meuller, B.O., Messing, M.E., Engberg, D.L., Jansson, A.M., Johansson, L.I., Norlén, S.M., Tureson, N., Deppert, K. (2012). Review of spark discharge generators for production of nanoparticle aerosols. *Aerosol Science Technolnology* 46:1256–1270.
[103] Michelsen, H. (2017). Probing soot formation, chemical and physical evolution, and oxidation: A review of in situ diagnostic techniques and needs. *Proceedings of the Combustion Institute* 36:717–735.
[104] Michelsen, H., Schulz, C., Smallwood, G., Will, S. (2015). Laser-induced incandescence: Particulate diagnostics for combustion, atmospheric, and industrial applications. *Progress in Energy and Combustion Science* 51:2–48.
[105] Motaung, D., Mhlongo, G., Kortidis, I., Nkosi, S., Malgas, G., Mwakikunga, B., Ray, S.S., Kiriakidis, G. (2013). Structural and optical properties of ZnO nanostructures grown by aerosol spray pyrolysis: Candidates for room temperature methane and hydrogen gas sensing. *Applied Surface Science* 279:142–149.
[106] Motzkus, C., Chivas-Joly, C., Guillaume, E., Ducourtieux, S., Saragoza, L., Lesenechal, D., Macé, T., Lopez-Cuesta, J.-M., Longuet, C. (2012). Aerosols emitted by the combustion of polymers containing nanoparticles. *Journal of Nanoparticle Research* 14:687.
[107] Mueller, R., Mädler, L., Pratsinis, S.E. (2003). Nanoparticle synthesis at high production rates by flame spray pyrolysis. *Chemical Engineering Science* 58:1969–1976.
[108] Nie, Y., Wang, Y., Biswas, P. (2017). Mobility and bipolar diffusion charging characteristics of crumpled reduced graphene oxide nanoparticles synthesized in a furnace aerosol reactor. *The Journal of Physical Chemistry C* 121:10529–10537.
[109] Oberreit, D.R., McMurry, P.H., Hogan Jr C.J. (2014). Mobility analysis of 2 nm to 11 nm aerosol particles with an aspirating drift tube ion mobility spectrometer. *Aerosol Science and Technology* 48:108–118.
[110] Oguri, K., Okano, Y., Nishikawa, T., Nakano, H. (2007). Dynamical study of femtosecond-laser-ablated liquid-aluminum nanoparticles using spatiotemporally resolved X-ray-absorption fine-structure spectroscopy. *Physical Review Letters* 99:165003.
[111] Okuyama, K., Lenggoro, I.W. (2003). Preparation of nanoparticles via spray route. *Chemical Engineering Journal* 58:537–547.
[112] Olfert, J. (2005). A numerical calculation of the transfer function of the fluted centrifugal particle mass analyzer. *Aerosol Science and Technology* 39:1002–1009.
[113] Olfert, J., Collings, N. (2005). New method for particle mass classification – The Couette centrifugal particle mass analyzer. *Journal of Aerosol Science* 36:1338–1352.

[114] Olfert, J., Reavell, K.S., Rushton, M., Collings, N. (2006). The experimental transfer function of the Couette centrifugal particle mass analyzer. *Journal of Aerosol Science* 37:1840–1852.

[115] Olfert, J., Symonds, J., Collings, N. (2007). The effective density and fractal dimension of particles emitted from a light-duty diesel vehicle with a diesel oxidation catalyst. *Journal of Aerosol Science* 38:69–82.

[116] Ong, P., Mahmood, S., Zhang, T., Lin, J., Ramanujan, R., Lee, P., Rawat, R. (2008). Synthesis of FeCo nanoparticles by pulsed laser deposition in a diffusion cloud chamber. *Applied Surface Science* 254:1909–1914.

[117] Ouyang, H., He, S., Larriba-Andaluz, C., Hogan, C.J. Jr (2015). IMS–MS and IMS–IMS investigation of the structure and stability of dimethylamine-sulfuric acid nanoclusters. *The Journal of Physical Chemistry. A* 119:2026–2036.

[118] Pan, Y., Tian, S., Liu, D., Fang, Y., Zhu, X., Zhang, Q., Zheng, B., Michalski, G., Wang, Y. (2016). Fossil fuel combustion-related emissions dominate atmospheric ammonia sources during severe haze episodes: Evidence from 15N-stable isotope in size-resolved aerosol ammonium. *Environmental Science & Technology* 50:8049–8056.

[119] Paraguay-Delgado, F., Antúnez-Flores, W., Miki-Yoshida, M., Aguilar-Elguezabal, A., Santiago, P., Diaz, R., Ascencio, J. (2005). Structural analysis and growing mechanisms for long SnO2 nanorods synthesized by spray pyrolysis. *Nanotechnology* 16:688.

[120] Park, S., Lee, K., Shimada, M., Okuyama, K. (2005). Coagulation of bipolarly charged ultrafine aerosol particles. *Journal of Aerosol Science* 36:830–845.

[121] Peineke, C., Attoui, M., Robles, R., Reber, A., Khanna, S., Schmidt-Ott, A. (2009). Production of equal sized atomic clusters by a hot wire. *Journal of Aerosol Science* 40:423–430.

[122] Peineke, C., Schmidt-Ott, A. (2008). Explanation of charged nanoparticle production from hot surfaces. *Journal of Aerosol Science* 39:244–252.

[123] Phanichphant, S., Liewhiran, C., Wetchakun, K., Wisitsoraat, A., Tuantranont, A. (2011). Flame-made Nb-doped TiO2 ethanol and acetone sensors. *Sensors* 11:472–484.

[124] Pratsinis, S.E. (1998). Flame aerosol synthesis of ceramic powders. *Progress in Energy and Combustion Science* 24:197–219.

[125] Rawat, V.K., Buckley, D.T., Kimoto, S., Lee, M.-H., Fukushima, N. (2016). Hogan Jr CJ: Two dimensional size–mass distribution function inversion from differential mobility analyzer–aerosol particle mass analyzer (DMA–APM) measurements. *Journal of Aerosol Science* 92:70–82.

[126] Reed, N., Fang, J., Chavalmane, S., Biswas, P. (2017). Real-time measurement of size-resolved elemental composition ratio for flame synthesized composite nanoparticle aggregates using a tandem SMPS-ICP-OES. *Aerosol Science and Technology* 51:311–316.

[127] Reischl, G.P., Mäkelä, J., Karch, R., Necid, J. (1996). Bipolar charging of ultrafine particles in the size range below 10 nm. *Journal of Aerosol Science* 27:931–949.

[128] Ren, Y., Li, S., Zhang, Y., Stephen, D.T., Long, M.B. (2015). Absorption-ablation-excitation mechanism of laser-cluster interactions in a nanoaerosol system. *Physical Review Letters* 114:093401.

[129] Ribeiro, J., Flores, D., Ward, C.R., Silva, L.F. (2010). Identification of nanominerals and nanoparticles in burning coal waste piles from Portugal. *Science of the Total Environment* 408:6032–6041.

[130] Rosser, S., de La Mora, J.F. (2005). Vienna-type DMA of high resolution and high flow rate. *Aerosol Science and Technology* 39:1191–1200.

[131] Saggese, C., Cuoci, A., Frassoldati, A., Ferrario, S., Camacho, J., Wang, H., Faravelli, T. (2016). Probe effects in soot sampling from a burner-stabilized stagnation flame. *Combustion and Flame* 167:184–197.

[132] Santos, J., Hontanón, E., Ramiro, E., Alonso, M. (2009). Performance evaluation of a high-resolution parallel-plate differential mobility analyzer. *Atmospheric Chemistry and Physics* 9:2419–2429.

[133] Seinfeld, J.H., Pandis, S.N. (2016). Atmospheric Chemistry and Physics: From Air Pollution to Climate Change, John Wiley & Sons, New York, US.

[134] Sethupathi, N., Thirunavukkarasu, P., Vidhya, V., Thangamuthu, R., Kiruthika, G., Perumal, K., Bajaj, H.C., Jayachandran, M. (2012). Deposition and optoelectronic properties of ITO (In 2 O 3: Sn) thin films by Jet nebulizer spray (JNS) pyrolysis technique. *Journal of Materials Science: Materials in Electronics* 23:1087–1093.

[135] Sgro, L., Barone, A., Commodo, M., D'Alessio, A., De Filippo, A., Lanzuolo, G., Minutolo, P. (2009). Measurement of nanoparticles of organic carbon in non-sooting flame conditions. *Proceedings of the Combustion Institute* 32:689–696.

[136] Sgro, L.A., D'Anna, A., Minutolo, P. (2010). Charge distribution of incipient flame-generated particles. *Aerosol Science Technolnology* 44:651–662.

[137] Sgro, L.A., D'Anna, A., Minutolo, P. (2011). Charge fraction distribution of nucleation mode particles: New insight on the particle formation mechanism. *Combustion and Flame* 158:1418–1425.

[138] Sharma, A.K., Rajaram, P. (2010). Nanocrystalline thin films of CuInS2 grown by spray pyrolysis. *Materials Science and Engineering: B* 172:37–42.

[139] Sharma, G., Wang, Y., Chakrabarty, R., Biswas, P. (2019). Modeling simultaneous coagulation and charging of nanoparticles at high temperatures using the method of moments. *Journal of Aerosol Science*, 132, 70–82.

[140] Sipkens, T., Mansmann, R., Daun, K., Petermann, N., Titantah, J., Karttunen, M., Wiggers, H., Dreier, T., Schulz, C. (2014). In situ nanoparticle size measurements of gas-borne silicon nanoparticles by time-resolved laser-induced incandescence. *Applied Physics B* 116:623–636.

[141] Sivakumar, P., Ramesh, R., Ramanand, A., Ponnusamy, S., Muthamizhchelvan, C. (2011). Preparation and properties of nickel ferrite (NiFe2O4) nanoparticles via sol–gel auto-combustion method. *Materials Research Bulletin* 46:2204–2207.

[142] Song, D.K., Dhaniyala, S. (2007). Nanoparticle cross-flow differential mobility analyzer (NCDMA): Theory and design. *Journal of Aerosol Science* 38:964–979.

[143] Steiner, G., Attoui, M., Wimmer, D., Reischl, G. (2010). A medium flow, high-resolution Vienna DMA running in recirculating mode. *Aerosol Science and Technology* 44:308–315.

[144] Stocker, T. (2014). Climate Change 2013: The Physical Science Basis: Working Group I Contribution to the Fifth Assessment Report of the Intergovernmental Panel on Climate Change, Cambridge University Press, Cambridge, UK.

[145] Stolzenburg, M.R., McMurry, P.H. (1991). An ultrafine aerosol condensation nucleus counter. *Aerosol Science and Technology* 14:48–65.

[146] Swihart, M.T. (2003). Vapor-phase synthesis of nanoparticles. *Current Opinion in Colloid & Interface Science* 8:127–133.

[147] Swihart, M.T., Catoire, L. (2000). Thermochemistry of aluminum species for combustion modeling from ab initio molecular orbital calculations. *Combustion and Flame* 121:210–222.

[148] Tang, Q., Cai, R., You, X., Jiang, J. (2017). Nascent soot particle size distributions down to 1 nm from a laminar premixed burner-stabilized stagnation ethylene flame. *Proceedings of the Combustion Institute* 36:993–1000.

[149] Thayer, D., Koehler, K., Marchese, A., Volckens, J. (2011). A personal, thermophoretic sampler for airborne nanoparticles. *Aerosol Science and Technology* 45:744–750.

[150] Timko, M.T., Yu, Z., Kroll, J., Jayne, J.T., Worsnop, D.R., Miake-Lye, R.C., Onasch, T.B., Liscinsky, D., Kirchstetter, T.W., Destaillats, H. (2009). Sampling artifacts from conductive silicone tubing. *Aerosol Science and Technology* 43:855–865.

[151] Tricoli, A., Elmøe, T.D. (2012). Flame spray pyrolysis synthesis and aerosol deposition of nanoparticle films. *AIChE Journal* 58:3578–3588.
[152] Tsantilis, S., Kammler, H., Pratsinis, S. (2002). Population balance modeling of flame synthesis of titania nanoparticles. *Chemical Engineering Science* 57:2139–2156.
[153] Tsantilis, S., Pratsinis, S.E. (2000). Evolution of primary and aggregate particle-size distributions by coagulation and sintering. *AIChE Journal* 46:407–415.
[154] Twomey, S. (1975). Comparison of constrained linear inversion and an iterative nonlinear algorithm applied to the indirect estimation of particle size distributions. *Journal of Computational Physics* 18:188–200.
[155] Ulrich, G., Milnes, B., Subramanian, N. (1976). Particle Growth in Flames, Experimental results for silica particles, II.
[156] Unni, H., Yang, C. (2005). Brownian dynamics simulation and experimental study of colloidal particle deposition in a microchannel flow. *Journal of Colloid and Interface Science* 291:28–36.
[157] Vander Wal, R.L., Dietrich, D.L. (1995). Laser-induced incandescence applied to droplet combustion. *Applied Optics* 34:1103–1107.
[158] Vander Wal, R.L., Ticich, T.M., West, J.R. (1999). Laser-induced incandescence applied to metal nanostructures. *Applied Optics* 38:5867–5879.
[159] Vanhanen, J., Mikkilä, J., Lehtipalo, K., Sipilä, M., Manninen, H., Siivola, E., Petäjä, T., Kulmala, M. (2011). Particle size magnifier for nano-CN detection. *Aerosol Science and Technology* 45:533–542.
[160] Vemury, S., Pratsinis, S.E., Kibbey, L. (1997). Electrically controlled flame synthesis of nanophase TiO 2, SiO 2, and SnO 2 powders. *Journal of Materials Research* 12:1031–1042.
[161] Wang, G., Zhang, R., Gomez, M.E., Yang, L., Zamora, M.L., Hu, M., Lin, Y., Peng, J., Guo, S., Meng, J. (2016). Persistent sulfate formation from London Fog to Chinese haze. *Proceedings of the National Academy of Sciences* 113:13630–13635.
[162] Wang, H. (2011). Formation of nascent soot and other condensed-phase materials in flames. *Proceedings of the Combustion Institute* 33:41–67.
[163] Wang, H., Zhao, H., Guo, Z., Zheng, C. (2012). Numerical simulation of particle capture process of fibrous filters using Lattice Boltzmann two-phase flow model. *Powder Technology* 227:111–122.
[164] Wang, J., Flagan, R.C., Seinfeld, J.H. (2002). Diffusional losses in particle sampling systems containing bends and elbows. *Journal of Aerosol Science* 33:843–857.
[165] Wang, Y.: Sub 2 nm Particle Characterization in Systems with Aerosol Formation and Growth, 2017.
[166] Wang, Y., Fang, J., Attoui, M., Chadha, T.S., Wang, W.-N., Biswas, P. (2014). Application of Half Mini DMA for sub 2 nm particle size distribution measurement in an electrospray and a flame aerosol reactor. *Journal of Aerosol Science* 71:52–64.
[167] Wang, Y., Kangasluoma, J., Attoui, M., Fang, J., Junninen, H., Kulmala, M., Petäjä, T., Biswas, P. (2017a). The high charge fraction of flame-generated particles in the size range below 3 nm measured by enhanced particle detectors. *Combust. Flame* 176:72–80.
[168] Wang, Y., Kangasluoma, J., Attoui, M., Fang, J., Junninen, H., Kulmala, M., Petäjä, T., Biswas, P. (2017b). Observation of incipient particle formation during flame synthesis by tandem differential mobility analysis-mass spectrometry (DMA-MS). *Proceedings of the Combustion Institute* 36:745–752.
[169] Wang, Y., Liu, P., Fang, J., Wang, W.-N., Biswas, P. (2015). Kinetics of sub-2 nm TiO2 particle formation in an aerosol reactor during thermal decomposition of titanium tetraisopropoxide. *Journal of Nanoparticle Research* 17:1–13.
[170] Wang, Y., Pinterich, T., Wang, J. (2018). Rapid measurement of sub-micrometer aerosol size distribution using a fast integrated mobility spectrometer. *Journal of Aerosol Science* 121:12–20.

[171] Wang, Y., Sharma, G., Koh, C., Kumar, V., Chakrabarty, R., Biswas, P. (2017c). Influence of flame-generated ions on the simultaneous charging and coagulation of nanoparticles during combustion. *Aerosol Science and Technology* 51:833–844.
[172] Will, S., Schraml, S., Leipertz, A. (1995). Two-dimensional soot-particle sizing by time-resolved laser-induced incandescence. *Optics Letters* 20:2342–2344.
[173] Wright, D.L., McGraw, R., Rosner, D.E. (2001). Bivariate extension of the quadrature method of moments for modeling simultaneous coagulation and sintering of particle populations. *Journal of Colloid and Interface Science* 236:242–251.
[174] Wu, J.J., Flagan, R.C. (1988). A discrete-sectional solution to the aerosol dynamic equation. *Journal of Colloid and Interface Science* 123:339–352.
[175] Xiong, G., Li, S., Zhang, Y., Buckley, S.G., Stephen, D.T. (2016). Phase-selective laser-induced breakdown spectroscopy of metal-oxide nanoparticle aerosols with secondary resonant excitation during flame synthesis. *Journal of Analytical Atomic Spectrometry* 31:482–491.
[176] Yan, W., Li, S., Zhang, Y., Yao, Q., Tse, S.D. (2010). Effects of dipole moment and temperature on the interaction dynamics of titania nanoparticles during agglomeration. *The Journal of Physical Chemistry C* 114:10755–10760.
[177] Yao, L., Garmash, O., Bianchi, F., Zheng, J., Yan, C., Kontkanen, J., Junninen, H., Mazon, S.B., Ehn, M., Paasonen, P. (2018). Atmospheric new particle formation from sulfuric acid and amines in a Chinese megacity. *Science* 361:278–281.
[178] Yatom, S., Bak, J., Khrabryi, A., Raitses, Y. (2017). Detection of nanoparticles in carbon arc discharge with laser-induced incandescence. *Carbon* 117:154–162.
[179] Yoon, C., McGraw, R. (2004a). Representation of generally mixed multivariate aerosols by the quadrature method of moments: I. Statistical foundation. *Journal of Aerosol Science* 35:561–576.
[180] Yoon, C., McGraw, R. (2004b). Representation of generally mixed multivariate aerosols by the quadrature method of moments: II. Aerosol dynamics. *Journal of Aerosol Science* 35:577–598.
[181] You, X., Wang, H., Goos, E., Sung, C.-J., Klippenstein, S.J. (2007). Reaction kinetics of CO+ HO2→ products: Ab initio transition state theory study with master equation modeling. *The Journal of Physical Chemistry. A* 111:4031–4042.
[182] Yu, M., Lin, J., Chan, T. (2008). A new moment method for solving the coagulation equation for particles in Brownian motion. *Aerosol Science and Technology* 42:705–713.
[183] Yuan, Y., Li, S., Yao, Q. (2015). Dynamic behavior of sodium release from pulverized coal combustion by phase-selective laser-induced breakdown spectroscopy. *Proceedings of the Combustion Institute* 35:2339–2346.
[184] Zachariah, M.R., Carrier, M.J. (1999). Molecular dynamics computation of gas-phase nanoparticle sintering: A comparison with phenomenological models. *Journal of Aerosol Science* 30:1139–1151.
[185] Zavodinsky, V., Mikhailenko, E. (2006). Quantum-mechanics simulation of carbon nanoclusters and their activities in reactions with molecular oxygen. *Computational Materials Science* 36:159–165.
[186] Zhang, H., Sharma, G., Wang, Y., Li, S., Biswas, P. (2019). Numerical modeling of the performance of high flow DMAs to classify sub-2 nm particles. *Aerosol Science and Technology* 53:106–118.
[187] Zhang, H., Swihart, M.T. (2007). Synthesis of tellurium dioxide nanoparticles by spray pyrolysis. *Chemistry of Materials* 19:1290–1301.
[188] Zhang, M., Wexler, A.S. (2006). Cross flow ion mobility spectrometry: Theory and initial prototype testing. *International Journal of Mass Spectrometry* 258:13–20.
[189] Zhang, S.-H., Flagan, R.C. (1996). Resolution of the radial differential mobility analyzer for ultrafine particles. *Journal of Aerosol Science* 27:1179–1200.

[190] Zhang, Y., Li, S., Ren, Y., Yao, Q., Law, C.K. (2014). Two-dimensional imaging of gas-to-particle transition in flames by laser-induced nanoplasmas. *Applied Physics Letters* 104:023115.
[191] Zhang, Y., Li, S., Ren, Y., Yao, Q., Stephen, D.T. (2015). A new diagnostic for volume fraction measurement of metal-oxide nanoparticles in flames using phase-selective laser-induced breakdown spectroscopy. *Proceedings of the Combustion Institute* 35:3681–3688.
[192] Zhang, Y., Li, S., Yan, W., Yao, Q., Tse, S.D. (2011). Role of dipole–dipole interaction on enhancing Brownian coagulation of charge-neutral nanoparticles in the free molecular regime. *The Journal of Chemical Physics* 134:084501.
[193] Zhang, Y., Wang, Z., Wu, X., Sun, L., Zhang, Z., Zhang, H., Li, S. (2017). In situ laser diagnostics of nanoparticle transport across stagnation plane in a counterflow flame. *Journal of Aerosol Science* 105:145–150.
[194] Zhang, Y., Xiong, G., Li, S., Dong, Z., Buckley, S.G., Stephen, D.T. (2013). Novel low-intensity phase-selective laser-induced breakdown spectroscopy of TiO2 nanoparticle aerosols during flame synthesis. *Combustion and Flame* 160:725–733.
[195] Zhao, B., Yang, Z., Wang, J., Johnston, M.V., Wang, H. (2003). Analysis of soot nanoparticles in a laminar premixed ethylene flame by scanning mobility particle sizer. *Aerosol Science and Technology* 37:611–620.
[196] Zhao, B., Zheng, H., Wang, S., Smith, K.R., Lu, X., Aunan, K., Gu, Y., Wang, Y., Ding, D., Xing, J. (2018). Change in household fuels dominates the decrease in PM2. 5 exposure and premature mortality in China in 2005–2015. *Proceedings of the National Academy of Sciences* 115:12401–12406.
[197] Zhao, H., Kruis, F.E., Zheng, C. (2010). A differentially weighted Monte Carlo method for two-component coagulation. *Journal of Computational Physics* 229:6931–6945.
[198] Zong, Y., Li, S., Niu, F., Yao, Q. (2015). Direct synthesis of supported palladium catalysts for methane combustion by stagnation swirl flame. *Proceedings of the Combustion Institute* 35:2249–2257.
[199] Stolzenburg, M.R., McMurry, P.H. (2008). Equations governing single and tandem DMA configurations and a new lognormal approximation to the transfer function. *Aerosol Science and Technology* 42(6):421–432.

Wei-Ning Wang and Xiang He

Chapter 3
Aerosol methodologies for synthesis of materials

Abstract: This chapter summarizes the representative aerosol methodologies for the synthesis of a variety of functional materials. The aerosol methodologies can be generally classified into three major categories: gas-to-solid conversion, liquid-to-solid conversion, and aerosol-assisted self-assembly. This chapter first describes the classical aerosol routes, such as chemical vapor deposition (CVD) and spray pyrolysis, by introducing the basic techniques and fundamental principles. The major focus is on the recently developed aerosol routes, such as aerosol CVD, salt-assisted spray pyrolysis, low-pressure spray pyrolysis, and evaporation-induced self-assembly. Strategies to control the particle size, particle size distribution, morphology, structure, composition, and functionality of the materials in those aerosol routes are discussed. Representative materials synthesized via the aerosol routes, such as particles, nanotubes, thin films, and hierarchical nanostructures, are introduced. Further, this chapter also reviews the recent progress on the aerosol processing of emerging materials, that is, crumpled graphene oxide nanoballs and metal-organic framework-based materials. The applications of the aerosol-processed materials in the sectors of energy, the environment, and human health are also briefly discussed.

Keywords: aerosol chemical vapor deposition, spray pyrolysis, spray drying, electrospray, evaporation-induced self-assembly, microdroplets, nanoparticles, thin films, crumpled graphene oxide, metal-organic frameworks, photocatalysts

---

Acknowledgments: The authors would like to acknowledge the financial support through the following funding agencies, including National Science Foundation (NSF), American Chemical Society Petroleum Research Fund (ACS-PRF), and Virginia Commonwealth University (VCU) through the Presidential Research Quest Fund (PeRQ) program.

Wei-Ning Wang, Department of Mechanical and Nuclear Engineering, Virginia Commonwealth University, 800 E. Leigh St., Richmond, VA 23219, USA, e-mail: wnwang@vcu.edu
Xiang He, Department of Mechanical and Nuclear Engineering, Virginia Commonwealth University, 800 E. Leigh St., Richmond, VA 23219, USA

https://doi.org/10.1515/9783110729481-003

3.1 Introduction

Aerosol processing of materials has been explored for several decades. A variety of aerosol methodologies have been developed by both academia and industry. In this chapter, several representative aerosol methodologies for the synthesis of materials are reviewed. In particular, the classification of aerosol methodologies is discussed, where the basic principles for the formation of functional materials via each aerosol method are explained, and the strategies to tune the material morphology, structure, composition, and functionality are introduced. The aerosol routes are also able to turn the two-dimensional (2D) graphene nanosheets into 3D crumpled nanoballs. These nanoballs not only resolve the restacking (or aggregation) issue of the 2D nanosheets but also could serve as a versatile platform to incorporate with other components to fabricate hybrid nanocomposites for energy, environmental, and medical applications. As an emerging porous material, metal-organic frameworks (MOFs) have attracted mounting attention during the past years. This chapter also introduces the facile synthesis of this novel material via a microdroplet-based aerosol approach. Microscopic heat and mass transfer associated with the MOF formation in the microdroplet approach is explained, and several important applications by using the MOF-based functional materials are demonstrated.

3.2 Classification of aerosol methodologies

Aerosol methodologies have been widely used for particle design and synthesis, which are characterized as simple, fast, continuous, scalable, and low cost with high production rates [9, 75, 106, 152]. The term "aerosol" was coined in the 1920s to describe a suspension of small particles in air or another gas [31, 50]. Therefore, the aerosol methodologies are traditionally classified as the gas-phase approaches. A typical example would be the chemical vapor deposition (CVD) method, where the materials are generated by chemical reactions of gaseous precursors [125]. In addition to the gas reactions, materials can also be formed through liquid-phase reactions in other aerosol processes, such as spray pyrolysis [106, 123], where the aerosolized droplets serve as the microreactors. Although solid materials can also be supplied as the precursors in high-temperature aerosol processes, such as plasma- and flame-assisted synthesis, the particle formation mechanisms are generally classified to be gas-to-solid conversion, since the solid materials are generally heated and vaporized as gaseous precursors in those processes. Therefore, in this chapter, the aerosol methodologies are classified into two major categories, that is, gas-to-solid conversion and liquid-to-solid conversion. As particles and droplets are the general forms of materials in many aerosol processes, the two categories are also frequently termed as "gas-to-particle conversion" and "droplet-to-particle conversion," respectively. Detailed particle formation mechanisms in each category are discussed in the following sections. In addition, to control the morphology and structure of materials via the aerosol routes, facile strategies based on the evaporation-induced self-assembly (EISA) have also been developed [10, 57, 92], which will be introduced in Section 3.2.3.

3.2.1 Gas-to-solid conversion

A gas-to-solid conversion process is a representative build-up method, where materials are generated by cooling a supersaturated vapor via either physical vapor deposition (PVD) or CVD. Both processes have been studied and reviewed extensively [103, 125]. This chapter first gives a brief introduction of the conventional thermal CVD process for particle synthesis, but puts more efforts toward the development progresses on the flame aerosol route and aerosol CVD (ACVD) method for nanostructured thin film deposition.

3.2.1.1 Particle synthesis

Particles are the general form of materials synthesized through a gas-to-solid conversion process. In a typical CVD process, a gaseous precursor is thermally decomposed or reacts with another precursor vapor. Particles are formed through a series of elementary steps, including but not limited to, nucleation, condensation, coagulation, and/or sintering. The development of strategies to efficiently control the particle size, particle size distribution, morphology, structure, and composition has been a major research focus. The key parameters include reaction temperature, process pressure, precursor feeding rate, overall gas flow rate, seed particles, and reactor dimensions [122, 124].

A typical experimental setup of a conventional thermal CVD process is schematically shown in Figure 3.1, which consists of a precursor feeding system, a gas flow control system, a furnace reactor, a Fourier-transform infrared detector to monitor the chemical compositions, and a corona discharge sampler to collect particles [111]. In Figure 3.1, the precursor feeding system is composed of a syringe pump and an evaporator to supply titanium tetraisopropoxide (TTIP) or titanium chloride ($TiCl_4$) vapor into the vertical electric furnace reactor where titanium dioxide (TiO_2) particles are formed through either thermal decomposition of the precursor vapor or hydrolysis reactions depending on the carrier gas type and the reaction temperature [111, 156]. As the heating source, besides the furnace reactor, various other means, such as a plasma chamber [78, 131], a laser-ablation system [29], or a flame burner can also be used [64, 126].

Taking a flame aerosol reactor (FLAR) as an example, a co-flow diffusion burner is typically used as the reactor [64]. In this method, organometallic precursors are

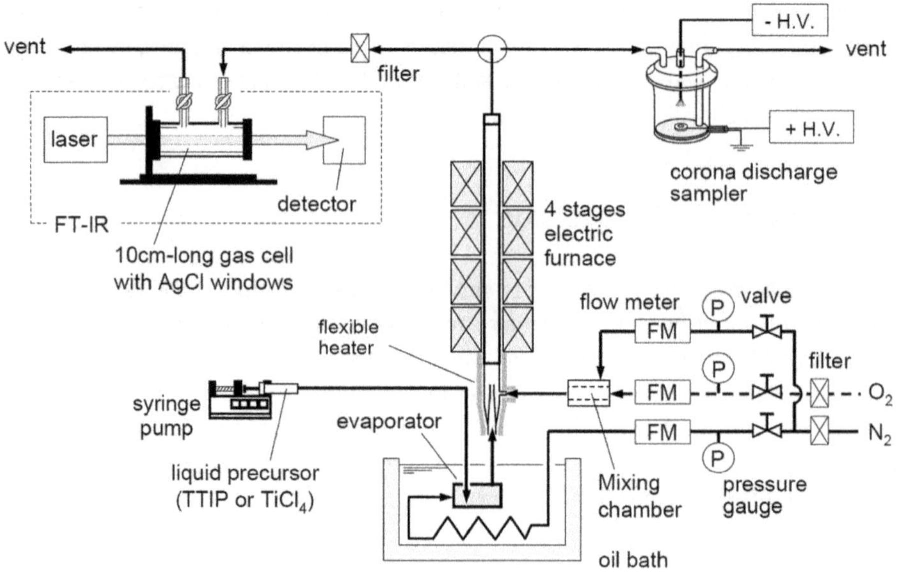

Figure 3.1: A thermal CVD process used to synthesize TiO_2 nanoparticles [111].

usually used, and the particle formation mechanism is generally considered based on the gas-phase reactions. The precursor is atomized into microdroplets by an atomizer or vaporized through a bubbler, which are then fed by the carrier gas into the flame zone, where the precursors undergo chemical reactions, nucleation, condensation, and growth to form the final particles. A quench ring may be used to control the flame length and temperature, the residence time, and hence the particle formation pathways during the synthesis by FLAR [64, 161]. Besides the co-flow diffusion burner, various other types of burners, such as counter-flow, McKenna flat-flame, Hencken flat-flame, and premixed burners, have also been designed and used in the flame aerosol synthesis as illustrated in Figure 3.2 [88]. The flame aerosol processes are widely used in both academia and industry to fabricate a variety of particles, including but not limited to ceramics, metals, metal oxides, nitrides, and carbides. The particle synthesis via the flame-assisted aerosol processes has been extensively reviewed elsewhere [88, 126, 134].

3.2.1.2 Nanotube synthesis

In addition to the particle formation, the gas-to-solid conversion processes are also used to synthesize one-dimensional (1D) materials, such as carbon nanotubes (CNTs). More specifically, CNTs are normally synthesized by using catalytic CVD, which involves several major steps: (1) deposition of catalyst on a

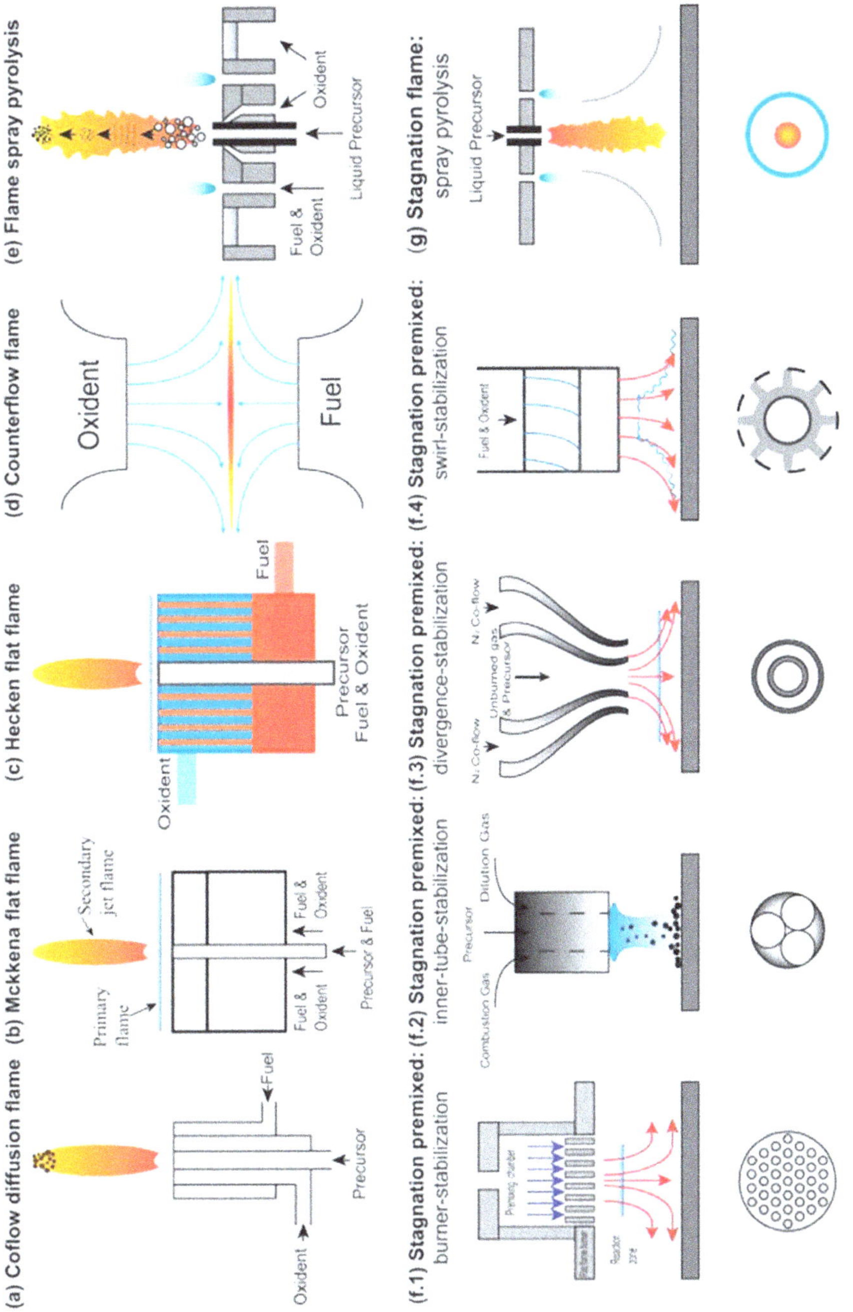

Figure 3.2: Schematic of various typical burner setups used in flame synthesis [88].

substrate; (2) introduction of hydrocarbon gas and carrier gas to the tube furnace; (3) decomposition of hydrocarbon vapor; (4) diffusion of carbon to catalyst; and (5) growth of CNTs after achieving the maximum solubility of carbon in the metal catalyst [82, 130]. There are two possible pathways: tip-growth model and base-growth model [82, 132]. The mechanism is closely related to the interaction between the catalyst and substrate. If the interaction is weak, CNT would come out across the bottom of the metal and continuously grow. Alternatively, CNT would grow upward from the top of the metal catalyst [5].

The synthetic parameters play a critical role in the production yield and quality of CNTs. For example, the purities and structural properties of CNTs are largely dependent on the choices of carbon source [107]. Compared with other hydrocarbon sources, ethanol is the most widely used precursor [102], as the as-prepared CNTs have high purity, which might be ascribed to the •OH-induced etching effect. Moreover, the addition of acetylene into ethanol would further retain the activity of the catalyst and accelerate the growth of CNTs [166]. Another interesting trend is that linear carbon sources (e.g., methane) normally produce aligned CNTs, while cyclic carbon sources (e.g., fullerene) often lead to curved CNTs [82]. Other parameters, such as catalyst species and reaction temperature, should also be given proper considerations for the synthesis of CNTs [8, 77].

3.2.1.3 Thin film deposition

A major application area of the gas-to-solid conversion processes is thin film deposition. For example, a FLAR system was developed to deposit nanostructured TiO_2 thin films with well-controlled morphologies [140–142]. As shown in Figure 3.3a, TTIP was fed by a bubbler to a premixed methane–oxygen (CH_4–O_2) burner, where the dimension of the burner, the gas flow rate, and the CH_4–O_2 ratio were adjusted to control the flame shape and temperature. In addition, the distance between the substrate and the burner head, and the substrate temperature were regulated to control the film deposition mechanisms. As illustrated in Figure 3.3b, the timescales for different processes, such as the chemical reaction of the precursor and aerosol dynamics, can affect the final film morphology [140]. If the characteristic time for the reaction of the precursor (τ_{rxn}) is larger than the residence time (τ_{res}), a CVD process is expected, where the TTIP molecules are transported to the substrate and react therein to form solid nanostructured TiO_2 film. Alternatively, if $\tau_{rxn} < \tau_{res}$, the precursor will react to form particles. Other characteristic times are also important to control the film morphologies, such as the particle–particle collision time (τ_{coll}) and the sintering time (τ_{sin}). Depending on the combinations of these timescales, the film deposition process would either be individual particle deposition or agglomerated particle deposition (Figure 3.3b). TiO_2 films with two representative morphologies can be obtained by the FLAR system: that is, granular films, which were deposited from the gas phase to form fractal structures and underwent little or no restructuring once deposited (Figure 3.4a), and columnar films, consisting of highly crystalline 1D structures oriented normal to the substrate (Figure 3.4b).

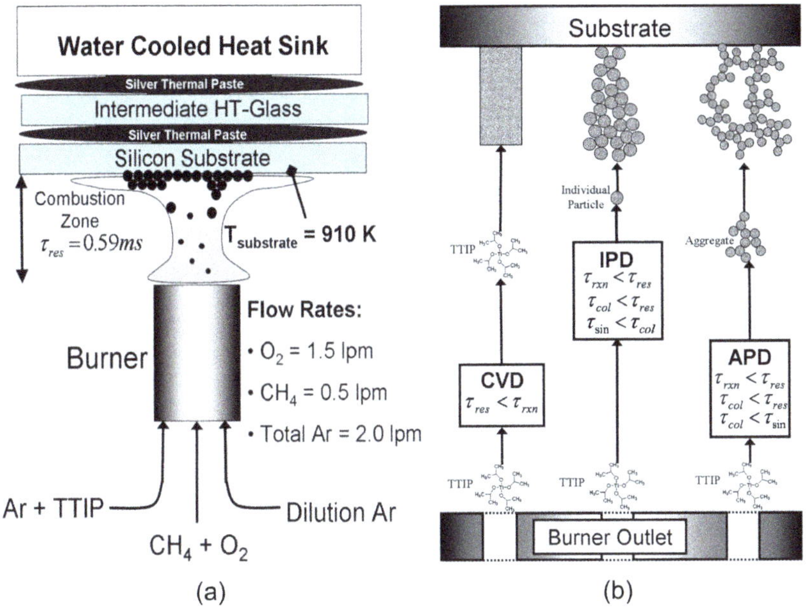

Figure 3.3: (a) Schematic diagram of the FLAR system for nanostructured film deposition; and (b) characteristic times for different film deposition processes [140].

In addition to the high-temperature FLAR system, an ACVD process was also developed to deposit nanostructured films [2–4, 38]. As illustrated in Figure 3.5, the combination of the residence time and reaction time determines the deposition and growth kinetics and hence the film morphology [3]. A follow-up research revealed that the competition between the sintering rate and the arrival rate is also a crucial factor. For a TiO_2 columnar film, the sintering time ($\tau_{\sin}(l)$) is defined as follows [2]:

$$\tau_{\sin}(l) = \frac{1.70 \times 10^{-3} kT(l) r_p{}^4}{b D_b \gamma \Omega} \tag{3.1}$$

where k is the Boltzmann constant, $T(l)$ is the temperature profile in a TiO_2 column which is a function of its length l, r_p is the radius of the particles, b is the boundary width of the grain, D_b is the diffusion coefficient of the grain boundary, γ is the surface tension, and Ω is the atomic volume.

To be specific, if the sintering rate is faster than the arrival rate of the deposited particles, a 1D columnar structure could be obtained. Further, the film thickness and structure were also dependent on the combination of the sintering rate and the

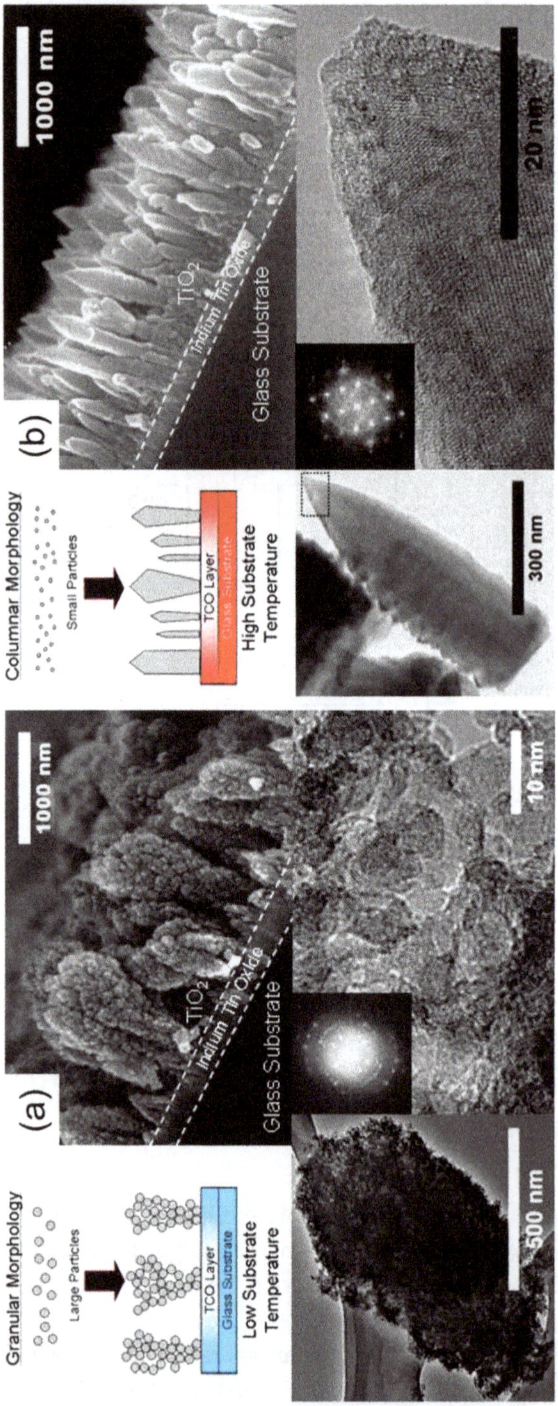

Figure 3.4: Nanostructured TiO_2 films with (a) granular and (b) columnar morphologies made by the FLAR process [142].

thermal resistance of the film [2]. In addition, a recent study by combining experiments with density functional theory calculations indicated that the preferential orientation of thin films is also a function of the system redox conditions [38].

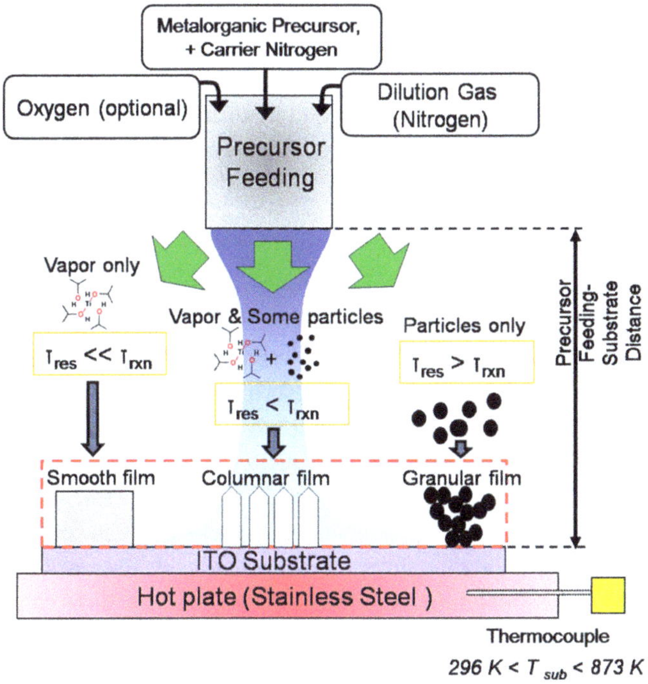

Figure 3.5: Illustration of the ACVD process with three deposition and film growth pathways [3].

In summary, the primary advantages of the gas-to-solid conversion aerosol routes are the controllable size, narrow size distribution, and high purity of the particles produced, enabling their applications for efficient synthesis of functional materials with controlled morphologies. However, as disadvantages, the formation of hard agglomerates in the gas phase leads to difficulties in preparing high-quality bulk materials. It is also challenging to synthesize multicomponent materials because of the differences in chemical reaction rates, vapor pressures, nucleation, and growth rates that occur during the gas-to-particle conversion process.

3.2.2 Liquid-to-solid conversion

A liquid-to-solid conversion process is often called a spray route, where the precursor liquid is atomized into microdroplets by an atomizer or nebulizer. These droplets then serve as microreactors for particle formation. An advantage of the spray routes over the gas-to-particle conversion routes is that multicomponent particles can be easily produced since homogeneous mixing of precursor components and stoichiometric reactions inside the microdroplets are feasible. The conventional spray route is often referred to as spray pyrolysis, where the precipitation, thermolysis, and sintering stages can be integrated into a single continuous process. The spray pyrolysis process has been extensively reviewed during the past decades [43, 75, 106, 123, 152]. In this chapter, the spray routes are classified into two major categories, that is, conventional spray routes and modified spray routes, including salt-assisted spray pyrolysis (SASP), low-pressure spray pyrolysis (LPSP), and chemical aerosol flow synthesis (CAFS).

3.2.2.1 Conventional spray routes

Conventional spray pyrolysis (CSP) in this chapter is specifically referred to as the spray pyrolysis process operated under atmospheric pressure, which uses an electric furnace as the heating source. In a CSP process, a starting solution is prepared by dissolving, usually, a metal salt in a solvent. The starting solution is atomized by an atomizer into droplets. These droplets are then introduced to the furnace reactor, where solvent evaporation, solute diffusion, drying, precipitation, reactions between the precursor and surrounding gas, pyrolysis, and/or sintering may occur inside the reactor to form final particles. By varying the droplet size, the furnace temperature, and the characteristic times of solvent evaporation (τ_{sv}) and solute diffusion (τ_{sl}), particles with various morphologies and sizes can be controlled and synthesized [123, 152].

Generally, the particle formation in the CSP process is governed by the one-droplet-to-one-particle (ODOP) conversion principle. For single-component particles, the correlation between the product particle diameter (d_p) and the droplet diameter (d_d) is described by eq. (3.2), assuming that the product particles are dense and spherical [152]:

$$d_p = d_d \left(\frac{C_d M_p}{\rho_p} \right)^{1/3} \tag{3.2}$$

where C_d, M_p, and ρ_p are the concentration of the precursor, the molecular weight, and the density of the product particles, respectively. For multicomponent particles, eq. (3.2) can be modified to the following equation [147]:

$$d_p = d_d \left(\frac{M_1 C_1}{\rho_1} + \frac{M_2 C_2}{\rho_2} \right)^{1/3} \tag{3.3}$$

where M_1, M_2, C_1, C_2, ρ_1, and ρ_2 are the molecular weights, concentrations, and densities of components 1 and 2, respectively.

The microscopic mass transfer (e.g., solvent evaporation rate) and heat transfer (i.e., droplet temperature variation) inside the microdroplets can be quantified based on the law of conservation of mass and energy. At pseudo-steady state, the evaporation rate from a single droplet is described by the following equation [84, 106]:

$$\frac{dm_d}{dt} = -\frac{4\pi r_d D_v M_g}{N_A}(n_s - n_g) \tag{3.4}$$

where m_d is the mass of a droplet, r_d is the droplet radius, D_v is the diffusion coefficient of solvent vapor, N_A is the Avogadro constant, M_g is the solvent molecular weight, and n_s and n_g are the vapor concentrations at the droplet surface and in the surrounding gas, respectively.

The temperature change in the droplet is given by eq. (3.5) by considering the energy balance between the droplets and surroundings:

$$\frac{dT_d}{dt} = \frac{1}{m_d C}\left[4\pi r_d K_g (T_g - T_s)\right] + \lambda\frac{dm_d}{dt} \tag{3.5}$$

where T_d is the droplet temperature, T_g is the surrounding gas temperature, C is the droplet heat capacity, K_g is the thermal conductivity of the surrounding gas, and λ is the latent heat of solvent evaporation.

Based on the above, the size of the final particles is mainly dependent on the droplet size and the precursor concentration. Commercial atomizers, such as a two-fluid nozzle or an ultrasonic nebulizer, are generally capable of generating droplets with an average size of several micrometers. Based on eqs. (3.2)–(3.5), most particles synthesized by a CSP process have sizes ranging from submicrometers to micrometers. The preparation of nanoparticles (d_d< 100 nm) via the CSP remains to be a challenge.

On the other hand, an electrohydrodynamic or electrospray pyrolysis (ESP) technique has demonstrated to be a viable method to atomize a liquid into ultrafine droplets [14, 15]. In a typical ESP process, the precursor solution is fed into a capillary tube and charged by a high-voltage power supply. The meniscus of the solution at the end of the capillary tube becomes conical where droplets are formed by the continuous breakup of a steady jet extending from the liquid cone, generally referred to as the "Taylor cone" [24]. Studies have shown that the diameter of the droplets may be controlled from nanometers up to millimeters [15, 129]. Theoretically, the scaling laws can be used to evaluate the size of the droplets generated in the ESP process [14, 25, 33]. These investigations revealed that the droplet initial diameter ($d_{d,i}$) is nearly independent of external electrostatic variables (e.g., electrode geometry and voltage), but scales with the electrical relaxation length (r^*) as follows [14]:

$$d_{d,i} = G(k)r^* \tag{3.6}$$

$$r^* = (Q\kappa\varepsilon_0/K_l)^{1/3} \tag{3.7}$$

where Q is the liquid flow rate, κ is the dielectric constant, ε_0 is the electrical permittivity of the vacuum, and K_l is the liquid electrical conductivity. Figure 3.6 shows the linear relationships between r^* and the diameter of droplets from eight solvents with different dielectric constants [14]. It is assumed that the particle formation in the ESP process is based on the ODOP principle. The product particle diameter can thus be predicted by eq. (3.2).

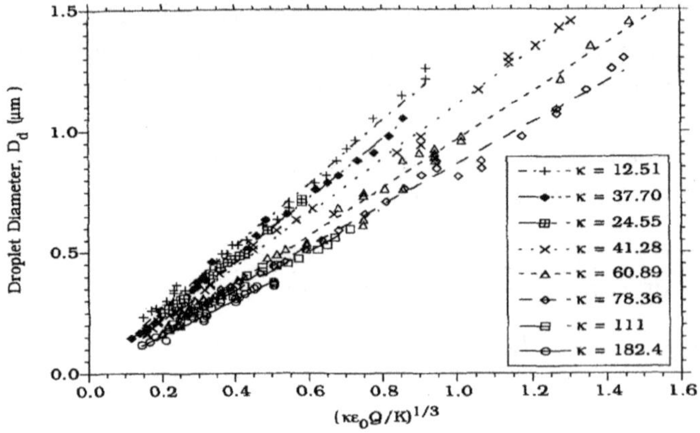

Figure 3.6: The linear relationship between droplet diameter and the electrical relaxation length [14].

3.2.2.2 Modified spray routes

As discussed above, except for the electrospray process, a CSP process can only produce submicrometer- to micrometer-sized particles due to the difficulty in generating ultrafine droplets (below 1 μm). Attempts have been made to modify the CSP process to produce nanoparticles. This chapter introduces several modified spray routes as shown below, where the particle formation mechanism is generally attributed to one droplet to multiple particles conversion.

3.2.2.2.1 Salt-assisted spray pyrolysis

In general, a particle synthesized by a CSP process consists of multiple primary nanocrystals, which are virtually inseparable due to the formation of a 3D network [106]. In order to separate these primary crystals, a strategy called salt-assisted spray pyrolysis was proposed by introducing compounds (typically inorganic salts) that can be distributed on the surfaces of the primary nanocrystals to prevent agglomeration and are then removed easily [162]. During the SASP process, single or eutectic salts, for example, chlorides or nitrates of Li, Na, and K, are dissolved in an aqueous precursor solution. The aerosolized droplets containing the mixture of salt and precursor are carried through a hot furnace where evaporation, nucleation, pyrolysis, and drying occur to form the final particles. Figure 3.7 shows two representative Y_2O_3–ZrO_2 particles prepared by CSP and SASP, respectively [162]. Both particles have a spherical shape after being collected (Figure 3.7a, c). However, after washing with water to remove the salts, the particles obtained from the SASP process disintegrated into nanocrystals (Figure 3.7d), while the CSP-generated particles were inseparable (Figure 3.7b). The separation of the primary nanocrystals is strongly dependent on the synthesis temperature. In general, the temperature should be higher than the melting point of a single salt or the eutectic temperature of the salt mixture to bring the salts to the liquid state in the microreactors functioning as high-temperature solvents [59, 164]. For this reason, SASP is also called a molten salt/flux method [98, 99]. The SASP method is universal, which can be used for preparing a variety of materials, such as Ni, NiO, CeO_2, Ag–Pd, CdS, ZnS, $LiCoO_2$, $BaTiO_3$, and GaN [,60–63, 121, 162–165].

Figure 3.7: Y_2O_3–ZrO_2 particles synthesized by CSP (a, b) and SASP (c, d) [162].

3.2.2.2.2 Low-pressure spray pyrolysis

Although the SASP process is efficient in producing various nanoparticles, it requires washing, drying, and salt recycling steps. To synthesize nanoparticles in a single step by using the solution-based aerosol process, an LPSP method was developed [71, 149]. The experimental setup of an LPSP process is similar to that of a CSP process, except that the system is maintained at low-pressure conditions, typically in the range of 10–80 torr (1,300–11,000 Pa). Various types of nanoparticles, such as ZnO [71], Ni [149], NiO [85], In_2O_3:Sn [119, 120], $BaTiO_3$ [153, 154], ZnO [49], and $LiCoO_2$ [48], were synthesized by the LPSP process. Example nanoparticles prepared by the LPSP process are shown in Figure 3.8a, b, where the particle size was mainly influenced by the pressure and gas flow rate [149]. In order to further understand the mechanism under the low-pressure environment, dispersion and aggregation of nanocolloids in microdroplets were also studied, where the rapid drying kinetics played a key role [151]. Theoretical studies on the evaporative cooling effect on nucleation and crystal growth were investigated as well. Under low-pressure conditions, the change in droplet size can be described as follows [30]:

$$\frac{dr_d(z)}{dz} = \frac{m}{u\rho_l}\left[\frac{p_v(T_m)}{\sqrt{2\pi mkT_m}} - \frac{p_s(T_d)}{\sqrt{2\pi mkT_d}}\right] \tag{3.8}$$

where u is the flow velocity, ρ_l is the liquid density, m is the mass of the vapor molecule, $p_v(T_m)$ is the partial water vapor pressure, $p_s(T_d)$ is the saturated vapor pressure, T_m is the mixture temperature, and $z$is the spatial variable. While the variation in droplet temperature, $T_d(z)$ is governed by the following equation [30]:

$$\frac{dT_d(z)}{dz} = \frac{3}{u\rho_l r_d(z)C\sqrt{2\pi k}}\{A\} - \frac{3T_d}{r_d}\frac{dr_d(z)}{dz} \tag{3.9}$$

where the explicit expression for A can be found in [30].

In an LPSP process, the microdroplets undergo rapid solvent evaporation upon entering the low-pressure chamber. The loss of latent heat from the droplets due to evaporation will cause a significant decrease in both the size and temperature of the droplets, the so-called evaporation cooling effect (see Figure 3.8c, d) [30]. The decrease in droplet and chamber temperatures is considered as a major promoting factor for the rapid nucleation and crystal growth inside the microdroplet. The agglomeration of primary nanocrystals will be limited due to the very short residence time under low-pressure conditions.

3.2.2.2.3 Other modified spray routes

In addition to the above methods, several other modified spray routes have also been developed in order to produce particles with controllable size, morphology, and composition. For example, Didenko and Suslick developed a chemical aerosol flow synthesis approach for the production of nanomaterials [27]. In this process, solutions of high-boiling-point liquids containing appropriate precursors are diluted with a low-boiling-point solvent (i.e., toluene). An aerosol is created using an ultrasonic nebulizer and the mist is carried out into a hot furnace maintained at 180–400 °C. As the droplet temperature exceeds the boiling point of toluene, the toluene evaporates, leaving submicron droplets of a concentrated solution of reactants in the high-boiling-point solvent. As the temperature inside the droplets further increases, the mixture inside reacts, forming the surfactant-coated nanoparticles. These nanoparticles are then collected and washed to obtain nanoparticles. This process is very similar to the SASP process, except for the solvents/flux used.

Adding polymer and organic additives into the precursor solution was also explored in the spray routes [70, 139, 150]. The key step is to add polymer and organic additives, such as polyethylene glycol, ethylene glycol, citric acid, and urea, into

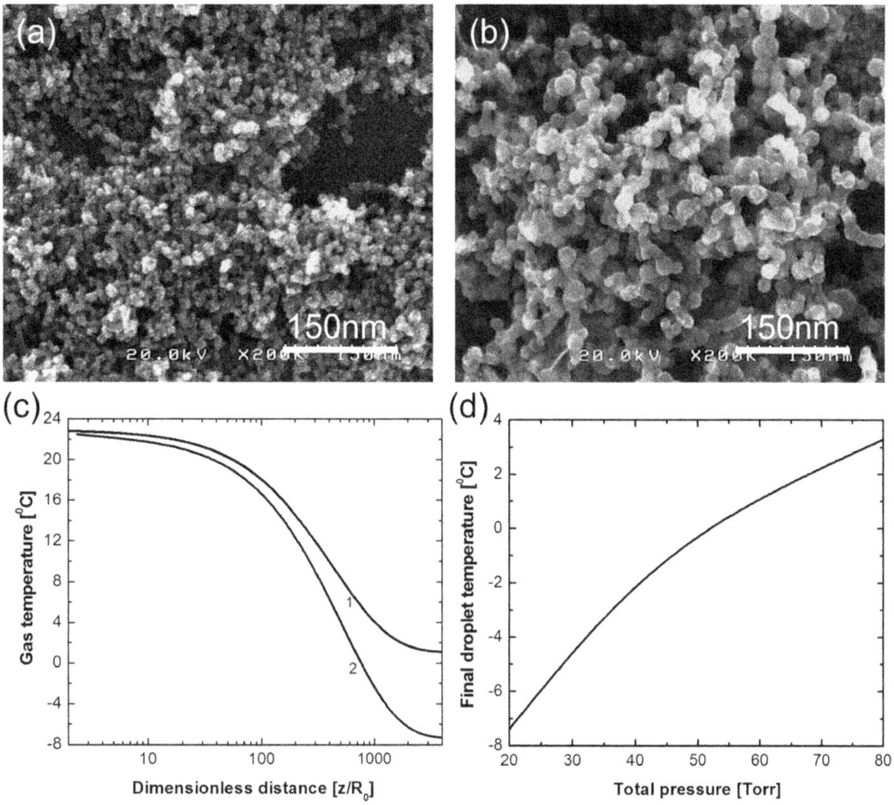

Figure 3.8: SEM images of Ni/NiO nanoparticles produced from $Ni(NO_3)_2$ at 1,100 °C and (a) 40 torr, 2 lpm, and (b) 80 torr, 2 lpm [149]. Simulated temperatures of (c) gaseous mixture and (d) droplets under low-pressure conditions [30].

the precursor solution. The as-prepared particles are in submicrometer with organic coatings. Post-annealing treatments at elevated temperatures are usually required to remove these coatings to disintegrate the 3D network to obtain nanocrystals. The additional combustion heat and evolution of gases during the thermal decomposition of the polymer and organic additives appear to be the main reasons for the fragmentation.

Various other technical modifications on the spray routes have also been made, such as the modification of ultrasonic nebulizers to produce monodisperse ultrafine droplets [74], development of a pneumatically assisted spray method [1], and use of a pulse combustor [160].

3.2.3 Aerosol-assisted self-assembly

The aforementioned aerosol methodologies, both gas-to-particle conversion and droplet-to-particle conversion, are capable of synthesizing particles with controlled sizes and shapes. In some cases, it is also desirable to design and fabricate particles with hierarchical structures at the microscales and nanoscales, such as porous, hollow, and other nanostructures, which have potential applications in catalysis, drug delivery, the environment, and energy conversion. In general, most of the above hierarchical nanostructures are made by self-assembly of certain building blocks, where the spontaneous organization of the materials is achieved through noncovalent interactions, such as hydrogen bonding, van der Waals forces, and electrostatic forces, with no external intervention [10]

Brinker and coworkers are the pioneers who first demonstrated that an aerosol-assisted self-assembly approach is effective in achieving mesoporous microspheres of silica [10, 92]. In this approach, a homogeneous solution of the mixture of soluble silica and a surfactant was prepared in high vapor pressure solvents, where the initial surfactant concentration (c_0) was much less than the critical micelle concentration (CMC). The precursor solution was atomized into microdroplets which were carried into a furnace, where the evaporation of the solvents induced micelle formation and successive co-assembly of silica-surfactant micellar species and further organization into crystalline mesophases. This process was termed as evaporation-induced self-assembly [92]. The transmission electron microscope (TEM) images of several representative mesoporous silica particles obtained by the EISA process are shown in Figure 3.9, where the mesoporous structures were controlled by the type and concentration of surfactants as well as the precursor components.

Okuyama and coworkers also used a spray approach to produce mesoporous particles [57, 58, 112, 113]. As illustrated in Figure 3.10, in this method, polystyrene (PS) latex spheres were used as templates, which were mixed with silica colloids in water to create a homogeneous precursor suspension. The suspension was sprayed into a furnace, consisting of two heating zones. The first zone was maintained at 200 °C to evaporate the solvent to produce composite particles consisting of primary silica particles and PS latex particles. The second zone was fixed at 450 °C to burn the PS latex particles, which resulted in the formation of porous silica particles in a single step. Theoretical analysis indicates that the hydrodynamic effects, such as microcirculation of the particles inside the droplet due to thermophoretic force and Brownian motion, during the drying process also play important roles in controlling the morphology of the resulting particles [56].

By manipulating the colloidal components, the size, weight percentage, and surface chemistry of the PS latex spheres, the ordered nanostructures of the final particles can be tailored. As shown in Figure 3.11, three different approaches can be

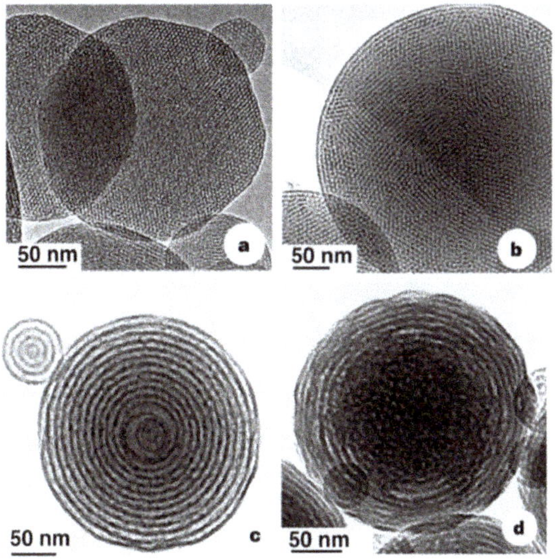

Figure 3.9: TEM images of mesoporous silica: (a) faceted calcined particles with a hexagonal mesophase with 5 wt% cetrimonium bromide (CTAB), (b) calcined particles with cubic mesostructures with 4.2 wt% Brij-58, (c) calcined particles with a vesicular mesophase with 5% P123, and (d) uncalcined particles with 2.5% Brij-56 [92].

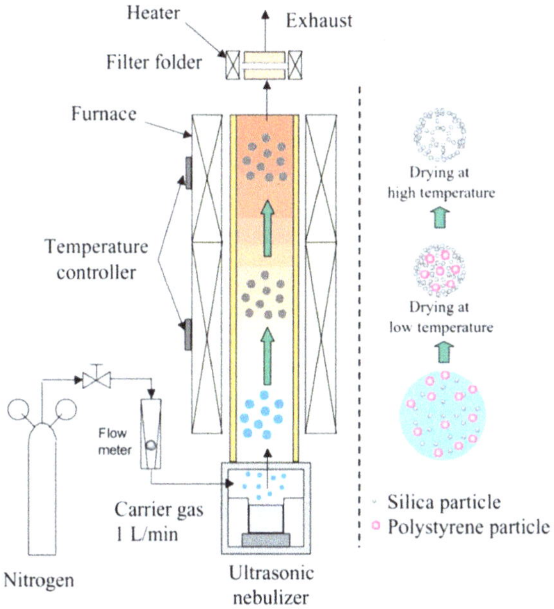

Figure 3.10: Schematic diagram of a spray drying setup where the PS spheres were used as the templates to synthesize porous particles in a single step [57].

used to produce silica particles with diverse nanostructures, that is, aggregated, porous, and hollow structures [83]. For example, using a controlled size separation procedure in aerosol routes, such as an aerosol impactor or a differential mobility analyzer, highly ordered aggregated particles with a fixed number of primary silica spheres with simple topological symmetry can be obtained.

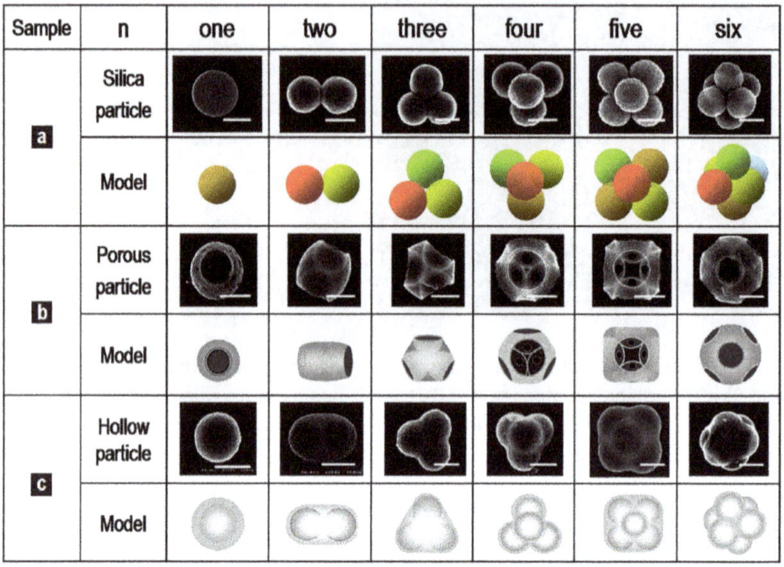

Sample
n
one
two
three
four
five
six
a
Silica particle
Model
b
Porous particle
Model
c
Hollow particle
Model

Figure 3.11: Scanning electron micrographs and schematic models of (a) aggregated particles of 100 nm silica for $n = 1–6$ and scale bars are 100 nm; (b) porous aggregates of 5 nm silica for $n = 1–6$ and scale bars are 150 nm, and (c) hollow aggregates of 5 nm silica for $n = 1–6$ and scale bars are 200 nm [83].

Recent studies also revealed that the electrostatic interactions between the PS latex spheres and the other precursor components play an important role in controlling the overall architecture of the resulting particles [6, 7, 83, 114, 115, 135]. For example, to prepare nanostructured carbon particles, a dual-polymer system comprising phenolic resin and charged PS latex spheres was studied via an ultrasonic spray pyrolysis method [7]. In this system, phenolic resin was selected as the carbon source, which is negatively charged due to the presence of OH^- groups. Both positively and negatively charged PS spheres were used as templates. As illustrated in Figure 3.12b, in general, hollow to porous carbon particles were formed when positively charged PS was used. In this case, the negatively charged phenolic resin was strongly bound to the positively charged PS spheres, packing the individual PS spheres tightly into the center of each droplet, resulting in the hollow structures. With the decrease in the magnitude of the surface charge of the PS templates, the hollow structures gradually changed to the porous structure. When negatively charged PS spheres were used, only macroporous carbon particles were formed since all components inside the droplets were experiencing electrical repulsive forces and kept separate throughout the droplets.

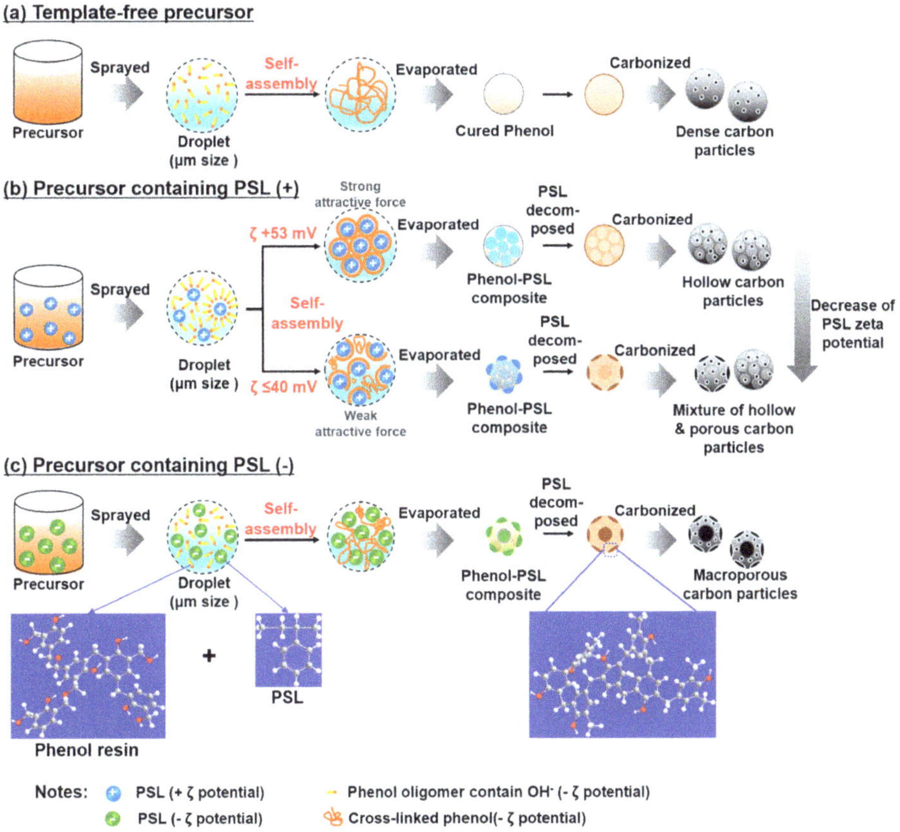

Figure 3.12: Schematic of nanostructure carbon formation by using charged PS templates [7].

3.3 Aerosol processing of emerging materials

In addition to the synthesis of conventional materials (e.g., polymers, silica, metals, and metal oxides), the aerosol methodologies also provide solutions to address issues existing with the emerging materials, such as restacking (or the so-called aggregation) of graphene nanosheets and slow nucleation and growth of MOFs. In this section, aerosol processing of the two representative emerging materials, that is, graphene and MOFs is summarized along with their applications.

3.3.1 Crumpled graphene oxide

Graphene, a single sheet of sp^2-hybridized carbon atoms, possesses many unique properties, including high surface area, extraordinary tensile strength, excellent thermal conductivity, and exceptional electron mobility [118]. These unique characteristics grant graphene great potentials in a myriad of applications, such as catalysis, energy storage, drug delivery, and biosensing [97, 127, 143, 157]. It should be noted that the graphene sheets have a strong tendency to aggregate and restack as a result of van der Waals attraction, which largely compromises the properties by decreasing surface areas. Strategies to address this issue have been developed, such as adding dispersing agents or decreasing the sheet size [94]. Another effective solution is to crumple the graphene, more practically, graphene oxide (GO) nanosheets into the ball structure via a microdroplet-based aerosol route, or the so-called spray route, which would not only maintain most of the advantages of graphene but also enable the samples to be resistant to compression and aggregation.

The evolution of GO structures during the aerosol process is shown in Figure 3.13. In this process, GO suspensions are first prepared by the modified Hummers' method [54], which are then atomized into microdroplets to be transported to a hot furnace, where the GO nanosheets deform under the evaporation-induced capillary force to be crumpled structures.

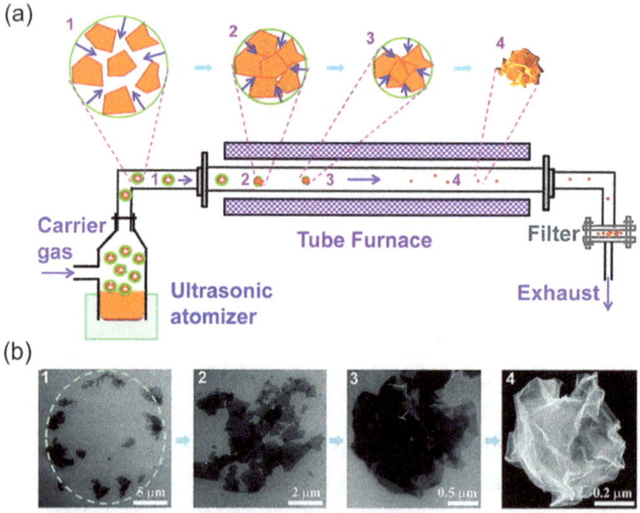

Figure 3.13: Crumpled GO sheets in evaporating aerosol droplets: (a) experimental setup and (b) SEM images of four samples collected along the flying pathway [94].

The online measurements of the crumpled GO (CGO) particles indicate that the fractal dimension of CGO varies from 2.54 to 2.68, depending on the solvent conditions [96].

Instead of thermal reduction, a capillary force is demonstrated to be the driving force for the crumpling process [72, 96, 145]. The capillary force is closely related to the evaporation rate of a solvent droplet. In general, a lower evaporation rate leads to a smaller capillary force, which would give rise to the formation of ripped GO rather than crumpled ones [145]. To quantify the capillary force, a universal equation from a single evaporating solvent droplet under steady state was developed [145]:

$$F = ABC_d^{\frac{1}{\delta D}} d_p^{-\frac{1}{\delta}} (\alpha T_d / P_d D_g)^{\frac{3}{\delta D}} \tag{3.10}$$

where F indicates the capillary force; A and B are used as constants; C_d implies the precursor concentration in the droplet; δ represents the force scaling exponent; D is fractal dimension; d_p is the particle diameter; α represents the solvent evaporation rate; T_d and P_d represent the droplet temperature and pressure, respectively; D_g indicates the gas-phase diffusion coefficient. As shown in Figure 3.14, the force increases with an increased evaporation rate, which results in smaller CGO. Combined with electron microscopy analysis, the critical capillary force to form CGO structures was determined to be ~30 μN [145].

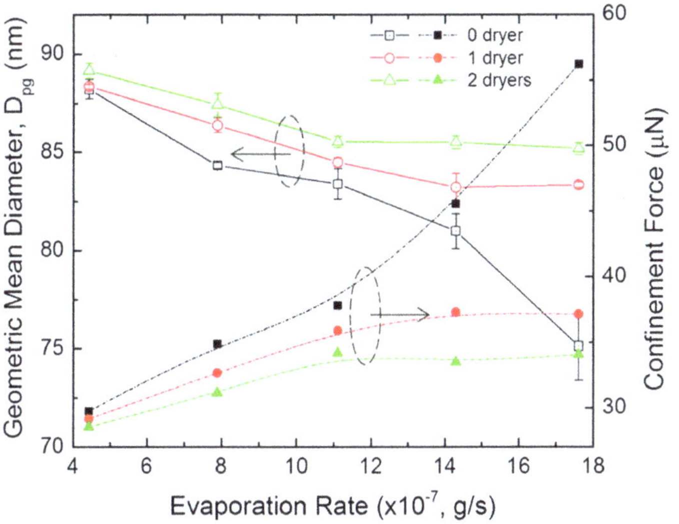

Figure 3.14: Geometric mean diameters of crumpled GO particles and the corresponding confinement force as a function of evaporation rate [145].

An additional benefit of the CGO structure is that it can serve as a carrier for various cargos (Figure 3.15a). The formation of these composites is dependent on the charges of GO and cargos [18]. In particular, pH is a critical parameter determining the surface charge (i.e., zeta potential) of the GO nanosheets in an aqueous suspension. As shown in Figure 3.15b, there are three electrostatic regimes: one attractive regime at moderate pH and two repulsive regimes at extreme pH. Specifically, GO and cargos are perfectly separated in the repulsive regime, which would result in good encapsulation of cargos in CGO (Figure 3.15c). In the attractive regime, cargos are already attached to GO sheets in the precursor solutions, which renders the appearance of cargos on the external surface of CGO after drying (Figure 3.15c). The carried cargos would extend the applications of graphene to different areas, including environmental sustainability, biomedical applications, and energy applications, which will be briefly described in the following sections.

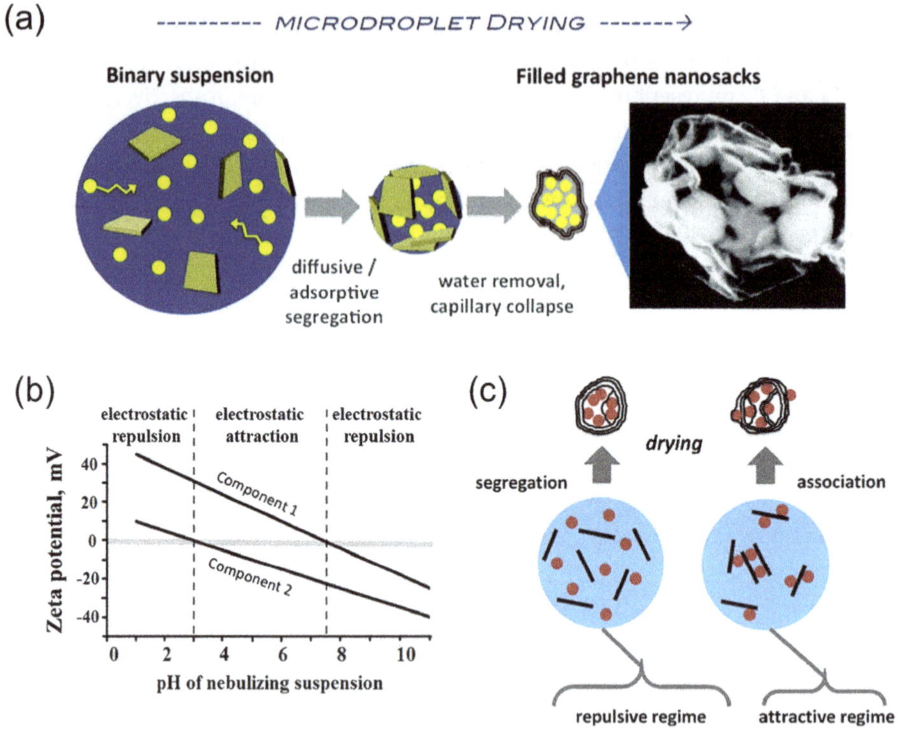

Figure 3.15: (a) Conceptual model for the colloidal self-assembly of filled graphene nanosacks [17]. Proposed theory for a colloidal sack–cargo assembly in binary systems: (b) pH-dependent zones of electrostatic attraction and repulsion, and (c) proposed assembly mechanisms in the repulsive and attractive regimes [18].

3.3.1.1 CGO for environmental sustainability

Given the high surface area and good electron mobility of CGO, massive studies have been carried out to fabricate CGO/catalyst composites as photocatalysts to combat environmental pollution issues, where CGO would greatly improve the photocatalytic efficiency by enhancing molecule adsorption and charge transfer. For example, a composite of TiO_2 encapsulated in CGO (hereafter CGO/TiO_2) was developed to reduce CO_2 to useful chemicals under irradiation of solar light [146, 173]. This process is often called CO_2 photoreduction, a promising approach to reduce the atmospheric CO_2 level and simultaneously convert CO_2 into value-added chemicals [89, 148, 155]. Compared with pure TiO_2, the CGO/TiO_2 photocatalyst exhibited significantly improved conversion efficiency. The dependence of the CGO/TiO_2 performance on the synthesis temperature was studied systematically. At a higher synthesis temperature, more functional groups were removed from the CGO surface, which created massive surface defects and impeded the electron mobility of CGO, making the CGO/TiO_2 composite less efficient as a photocatalyst. Instead, if synthesized at low temperatures (e.g., 200 °C), the CGO would preserve most of the surface functional groups and smooth surface. Upon light irradiation, the encapsulated TiO_2 would reduce the CGO by removing the functional groups without creating too many defects. The reduced CGO would act as the electron sink and significantly enhance the photocatalytic efficiency by increasing the lifetimes of charge carriers. The photocatalytic performance of CGO/TiO_2 can be further improved by adding amine groups onto CGO [117]. In addition to CO_2 photoreduction, CGO/TiO_2 composite is also an excellent photocatalyst to degrade organic pollutants in water through both reductive and oxidative reactions [66]. Meanwhile, by adding magnet particles in the composite, the catalysts can be easily recovered from the aqueous environment [66].

Additionally, the CGO-based composites are also excellent candidates to fabricate advanced membrane filtration systems to tackle water contamination issues [65, 67]. For instance, Jiang et al. designed a multifunctional membrane with CGO/TiO_2/Ag deposited on the poly(ether sulfone) support [67]. The deposited CGO-based composite was stabilized with polyallylamine by forming C–N bonds with the CGO. The as-fabricated membrane not only shows high permeability because of the channels formed by CGO but also exhibits several other attracting features, such as high rejection rate, photocatalytic reactivity, and antimicrobial ability. The antimicrobial activity of the membrane primarily was attributed to the presence of silver (Ag) or the dissolved silver ions (Ag^+), which could cause damages to the cell membrane or DNA [174]. However, with prolonged functioning time, the Ag would gradually dissolve in the solution as Ag^+, which is detrimental to maintain the efficiency of the membrane in terms of disinfection. To address this issue, Jiang et al. later advanced their membrane technology by regenerating Ag nanoscale particles on the surface of CGO/TiO_2 membrane via photocatalytic reduction of the dissolved Ag^+ [65].

3.3.1.2 CGO for biomedical applications

The outstanding features of graphene and graphene-derived materials also grant them huge potentials in the biological and medical applications [143]. For instance, the integration of GO with silk fibroin would lead to intermolecular forces between these two components, which significantly improved biocompatibility, mechanical strength, and stability, making this composite suitable for cell adhesion and proliferation [144]. The intrinsic optical properties of GO lead to them being widely used in biosensing and photothermal therapy [143]. For example, Robinson et al. designed PEGylated rGO sheets, which are stable and exhibit strong near-infrared absorption ability, allowing the conjugation of a peptide for the targeting and photothermal ablation of cancer cells [128]. Due to their large surface areas and abilities to bind drugs through either hydrophobic or π–π interactions, graphene derivatives have also been used for drug delivery [109].

It should be noted that the aforementioned graphene-based hybrid materials typically demand complicated synthetic procedures. The aerosol processes appear to be an efficient approach, which allows facile fabrication of CGO with mesoporous channels. These open channels create opportunities for the controlled release of the loaded drugs. For example, Chen et al. fabricated the CGO/CsCl/CMC (CMC: carboxymethyl cellulose, Figure 3.16a) using the aerosol process [18]. The CsCl component can be released through the channels created by CGO to inhibit the growth of tumors, while the CMC was used as a barrier to control the release rate of CsCl. As shown in Figure 3.16b, without CMC, the release of CsCl is completed within 10 s. The addition of CMC significantly reduced the release rate of CsCl, making it possible for controlled drug release (Figure 3.16b).

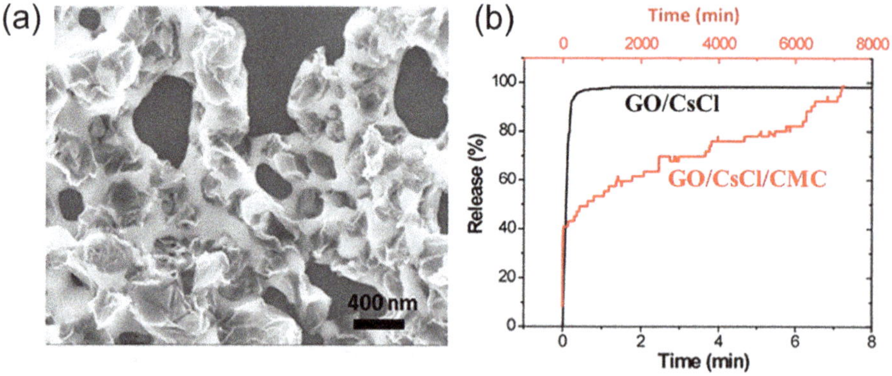

Figure 3.16: (a) SEM of CsCl-filled graphene nanosacks after brief exposure to humid air. (b) Chloride release rate in DI water for CsCl-filled nanosacks with/without CMC [18].

When it comes to biological or medical applications, the first concern of the nanomaterials is their biocompatibility and toxicity. Due to variations in experimental conditions, the influence of graphene-derived biomaterials on the biological system is still a matter of debate [170]. In general, the pristine graphene and its derivatives are considered to be toxic with a dependence on dose [169]. However, their toxicity can be greatly reduced via functionalization by the biocompatible polymer [28].

3.3.1.3 CGO for energy application

(1) Lithium-ion batteries (LIBs). Given their high theoretical capacities, metal oxides (e.g., NiO and Fe_3O_4) and group IVA elements (e.g., Si) are widely adopted as the anode materials for lithium-ion batteries (LIBs) [20, 95]. However, most of these materials undergo capacity fading upon lithiation/delithiation, which is mainly due to the material fracture caused by drastic volume fluctuation [159] and deposition of solid electrolyte interphase layers [138]. Recently, the integration of these anode materials with graphene sheets was conducted to address the above issues, which are generally synthesized via either assembly of anode materials with graphene or in situ growth of anode materials on graphene [167].

In comparison, the aerosol process not only simplified the procedures but also created intriguing crumpled structures. Kang and coworkers synthesized several CGO-based anode materials using the aerosol process, such as rGO/V_2O_5 [22], rGO/GeO_x [19], rGO/Ni/NiO [23], rGO/MoO_3 [20], and rGO/Fe_3O_4 [21]. With the integration of rGO, these composites exhibit superior electrochemical properties than the bare metal oxides. Taking rGO/molybdenum oxide for an example (Figure 3.17a), the comparison between bare MoO_3 and rGO/MoO_3 in terms of electrochemical properties is shown in Figure 3.17b–e. In particular, the incorporation of the crumpled rGO improved the initial discharge capacity (current density: 2 A/g) from 1,225 to 1,490 mAh/g, meanwhile, increased the capacity retention from 47% to 87% [20]. Liu et al. demonstrated that CGO balls could serve as promising building blocks to stabilize Li metal anodes, preventing them from dendritic growth [91], which resulted in a huge coulombic efficiency (i.e., 97.5%) with extraordinary stability (>750 cycles or 1,500 h).

(2) Supercapacitors. The CGO is also a preferred material for supercapacitors than the 2D graphene sheets in terms of power density and cycle life. In particular, the porous channels in CGO could improve electrical conductivity and electrolyte accessibility, and the crumpled ball structure could overcome restacking issues that could largely impede the transportation of ions and electrons [11, 101, 116]. As a demonstration, Luo et al. compared the performance of graphene in three morphologies (i.e., flat, wrinkled, and crumpled) as the capacitor materials (Figure 3.18) [93]. The results showed that all three graphene materials behaved as an electrical double-layer capacitor electrode with coulombic efficiency close to 100%. While the specific capacity of CGO outperforms those of flat and wrinkled graphene sheets even with high mass loading and large current density (Figure 3.18d–e) [93]. A later study by Jo et al. showed that the capacitive performance of CGO balls was highly dependent on their size and structure [68]. In particular, lower resistance for charge transfer could be achieved with larger CGO balls, and the ion transport would be facilitated with more meso- and microporous structures.

The mass transport among CGO materials can be further improved by decorating CNTs on the CGO surface [100]. As a result, this hybrid exhibited remarkably improved specific capacitance, rate capabilities, and energy density. In addition, polyaniline

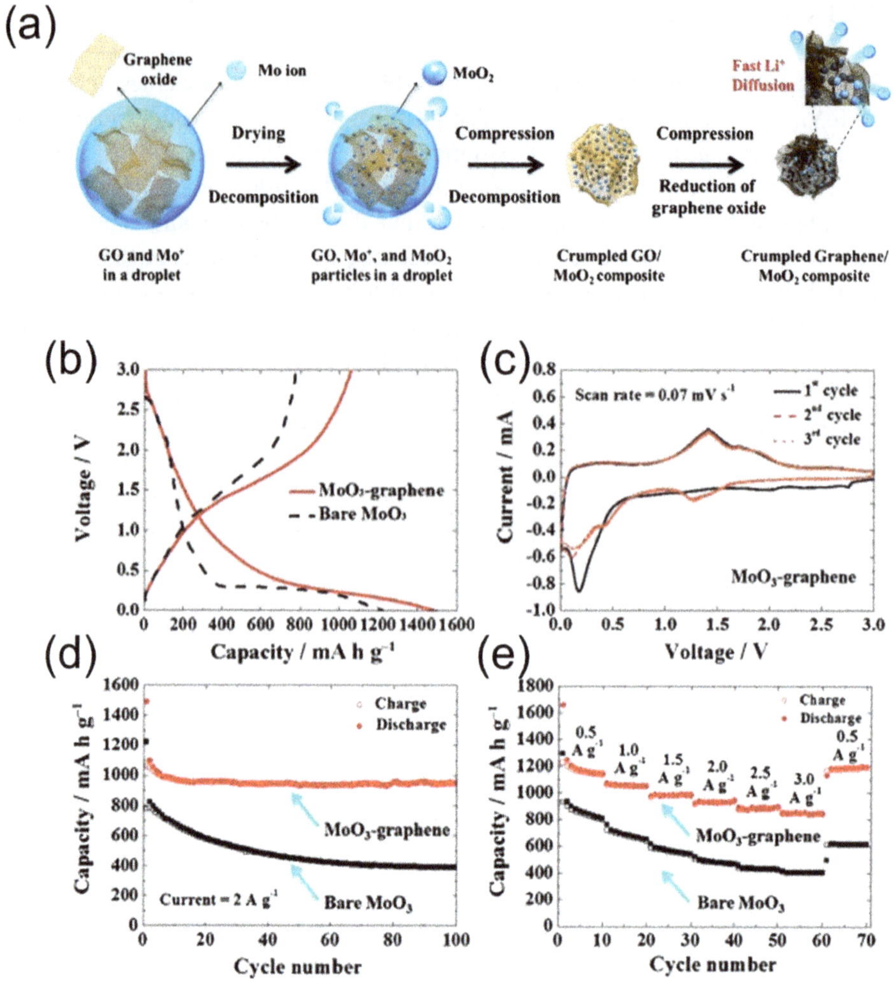

Figure 3.17: (a) Schematic diagram of the formation mechanism of the crumpled graphene–MoO_2 composite powders by an aerosol process; electrochemical properties of the crumpled graphene–MoO_3 composite and bare MoO_3 powders: (b) initial charge/discharge curves, (c) CV curves, (d) cycle performances, and (e) rate performances [20].

could also be added into the composite to provide high pseudocapacitance [69]. It should be noted that conducting binders are required to fabricate CGO electrodes. Recently, a strategy to fabricate CGO electrodes without binders was developed by using an electrospray deposition [137]. In this process, an electric stove was used to create a heating zone underneath the spray region, where the evaporation-induced crumpling graphene sheets occurred. A stainless-steel substrate was used to collect the CGO film, which could directly serve as the capacitor electrode after the reduction treatment.

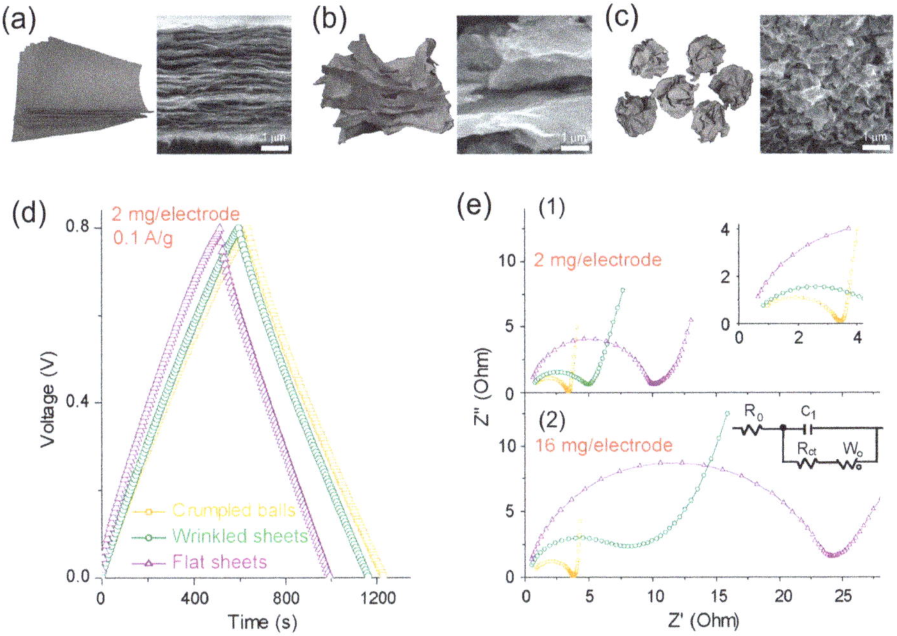

Figure 3.18: Paper models and SEM images showing stacks of (a) flat graphene sheets, (b) heavily wrinkled sheets, and (c) crumpled graphene balls. (d) Typical charge/discharge curves of symmetric ultracapacitor devices at 0.1 A/g. Nyquist plots of the three samples with mass loading of (e1) 2 mg and (e2) 16 mg per electrode. The inset of (e1) shows the zoom-in view of the intersections with the *Z′*-axis showing the ohmic resistance of the devices. The inset of (e2) shows an equivalent circuit of the devices [93].

3.3.2 Metal-organic frameworks

MOFs are highly porous materials built out of metal clusters and organic ligands through coordination bonds [76, 168]. Typical merits of MOFs include super high surface area, exceptional porosity, and chemical and structural tunability, giving them great potentials over a broad of areas [32], such as gas adsorption, separation and storage [46, 47, 86, 136], catalysis [26, 172], drug delivery [52, 55], and sensing [53, 79]. Given the affluence of metal and ligand species, a variety of MOF structures (>70,000) have been designed [108].

3.3.2.1 MOF particles

MOFs are conventionally synthesized by wet-chemistry methods, where nonuniform heat and mass transfer and batch-to-batch errors are consistent issues. An aerosol process appears to be an ideal approach to tackle these problems [12, 34, 39, 41, 44, 45, 81, 158]. As a demonstration, several typical MOFs have been synthesized by using the aerosol routes [12], including HKUST-1, MOF-5, MOF-14, Cu-BDC, ZIF-8, and UiO-66. The as-prepared MOF materials are generally in spherical shapes, assembled by massive nanosized MOF crystals. With easy sonication, the spherical samples would disassemble into homogenous nanocrystals (Figure 3.19a–c) [12].

The production yields and the physical properties of MOFs are controllable by simply adjusting the operating parameters of the aerosol process, including precursor conditions (e.g., concentration, pH, ratio, and additives), reactor temperature, and operating pressure, which are detailed in the following sections.

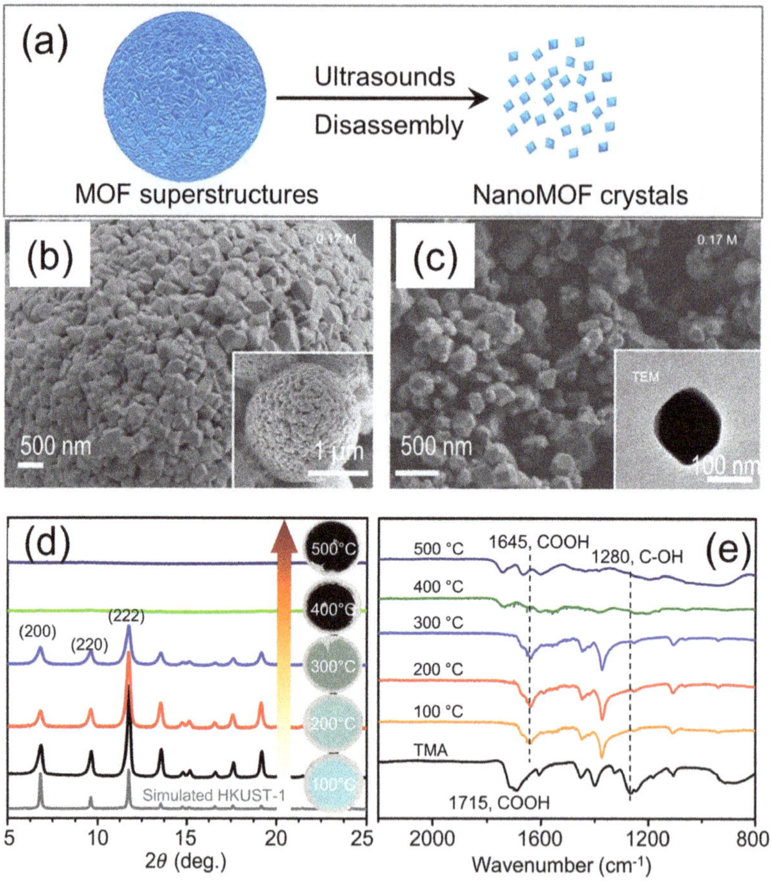

Figure 3.19: (a) Schematic showing the disassembly of the HKUST-1 superstructures on sonication to form well-dispersed, discrete nano-HKUST-1 crystals; representative FESEM and TEM (insets) images of the HKUST-1 superstructures (b) and corresponding disassembled nano-HKUST-1 crystals (c) [12]; (d) XRD patterns and (e) Fourier-transform infrared spectra of HKUST-1 samples synthesized at different temperatures [41].

Precursor parameters. Based on eq. (3.2), increased precursor concentrations would increase the production yields and sizes of the MOF crystals. Taking HKUST-1 as an example [12], with all other parameters fixed, increasing the precursor concentration from 0.0017 to 0.17 mol/L would increase the particle size and yield from 75 nm and 70% to 160 nm and 78%, respectively. The pH of the precursor solution also plays a significant role in the synthesis of MOFs. In general, a high pH value would promote the deprotonation of the ligands and expedite the nucleation process, which eventually leads to better production yields and textural properties [81].

Another strategy to tune the deprotonation rates of organic ligands is to adjust the ratio of the metal ions and ligands, that is, metal/ligand ratio in the precursor solution. As reported in a prior study [41], with high metal/ligand ratios, the excessive metal ions would also enhance the deprotonation rate of the ligand, which subsequently promotes the nucleation rates and gives rise to smaller crystal sizes. However, the metal/ligand ratio should be within a reasonable range to ensure the perfect MOF formation.

The functions of the MOFs could be further expanded by adding a second or third metal ion component or ligand component in the precursor solutions to synthesize mixed-linker or bimetallic MOFs [39, 45]. For instance, $Cu(NO_3)_2 \cdot 3H_2O$/TMA/BDC (TMA: trimesic acid; BDC: 1,4-benzenedioic acid) mixture and $Ni(NO_3)_2 \cdot 6H_2O/Zn(NO_3)_2 \cdot 6H_2O$/dhbdc (dhbdc: 2,5-dihydroxy-1,4-benzenedicarboxylate) mixture would produce Cu-TMA/BDC and Ni/Zn-MOF-74 superstructures, respectively [12, 45]. By adjusting the ratios of the precursor components, the properties (e.g., crystalline structure, textural properties, and surface chemistry) of the mixed-component MOFs are changeable. Interactions exist between the components, which were demonstrated with the aid of 2D correlation spectroscopy analysis [45].

Moreover, the functionality and stability of MOFs can also be easily improved with additives in the precursor solutions. For instance, the precursor solution can also be supplemented with amines or aldehydes for the covalent post-synthetic modification of MOFs via Schiff-base condensation reactions [36]. To protect hydrolytic unstable MOFs from collapsing, hydrophobic organic polymers can be added in the precursor solution of the aerosol process to generate a protection layer on the MOF surface [13]. In addition, CTAB could be added in the precursor, which will serve as a template to create hierarchical MOF structures [34].

Reactor temperature. With the increased reactor temperature, higher production yields would be achieved, which could be attributed to the higher nucleation and crystallization rates, and the reduced chances of particle losses inside the reaction chamber due to diffusive deposition [81]. In addition, the reactor temperature also plays a significant role in the sizes and dehydration degree of MOF products [41]. With other parameters controlled, the MOF particles synthesized at high temperatures generally have smaller sizes than those synthesized at low temperatures, which is due to the fact that nucleation is more favorable than the crystal growth for MOFs at high temperatures [41]. Besides, the variations in reactor temperature also have significant effects on the dehydration degree of the MOF products. Higher dehydration degree means more solvents have been released from the MOF structure, leaving massive coordinatively unsaturated metal sites, which are perfect for gas adsorption, molecule activation, and catalysis [42, 51, 171]. On the other hand, if the reactor temperature is over the limit of the MOFs, the MOF crystals would be destroyed [41]. As shown in Figure 3.19d–e, both the crystal structure and surface functional groups of HKUST-1 are damaged under temperatures over 300 °C.

Operating pressure. Compared with the reactor temperature, the operating pressure seems to be a milder parameter to be adjusted for the optimization of MOFs. By using Cu-BDC as a prototypical MOF, He and Wang demonstrated that the varying operating pressures would change the sizes and textural properties of MOFs (Figure 3.20) [44]. When synthesized at ambient pressure, the Cu-BDC has a sheet morphology with a length of 915 nm. With decreased operating pressures, the sizes of Cu-BDC gradually reduced to 299 nm. Besides, the low operating pressures could also lead to the release of solvent molecules from the metal sites, giving rise to changes in crystalline structure, textural properties, and gas sorption behaviors [44].

The dependence of sizes of MOFs on the reactor temperature and operating pressure is related to the evaporation of microdroplets during the spray process. In general, the high temperatures and low operating pressures would increase the evaporation rate of microdroplets, which would increase the supersaturation ratios and subsequently promote the nucleation process.

3.3.2.2 MOF thin films

In addition to the particle morphology, the aerosol process also enables the manufacturing of MOF thin films [73, 104, 105]. For instance, a gravimetric sensor was made by Khoshaman et al. by depositing MOF layers onto a quartz resonate using the electrospray process [73]. Compared to those fabricated using the drop-casting method, the as-prepared sensor exhibits better performance regarding resolution and stability. Similarly, Melgar et al. successfully fabricated ZIF-7 and ZIF-8 membranes on the α-alumina substrate by using the electrospray process for gas separation (Figure 3.21a) [104, 105]. The whole process completed in a fast manner, and the thickness of the ZIF membranes is controllable by adjusting the synthesis temperature and sprayed volume. The as-prepared membranes exhibited great H_2/CO_2 selectivity. The highest H_2/CO_2 selectivity of the as-prepared membrane was found to be ~20, which is more than 4 times of the H_2/CO_2 Knudsen separation factor (i.e., 4.7), demonstrating the molecular sieving effects of the products. Another way to synthesize MOF thin films is to spray droplets containing metal ions to the organic ligand solutions [90]. Due to anisotropic diffusion, the MOF nanosheets will be formed laterally at the interface between the metal solutions and organic ligand solutions.

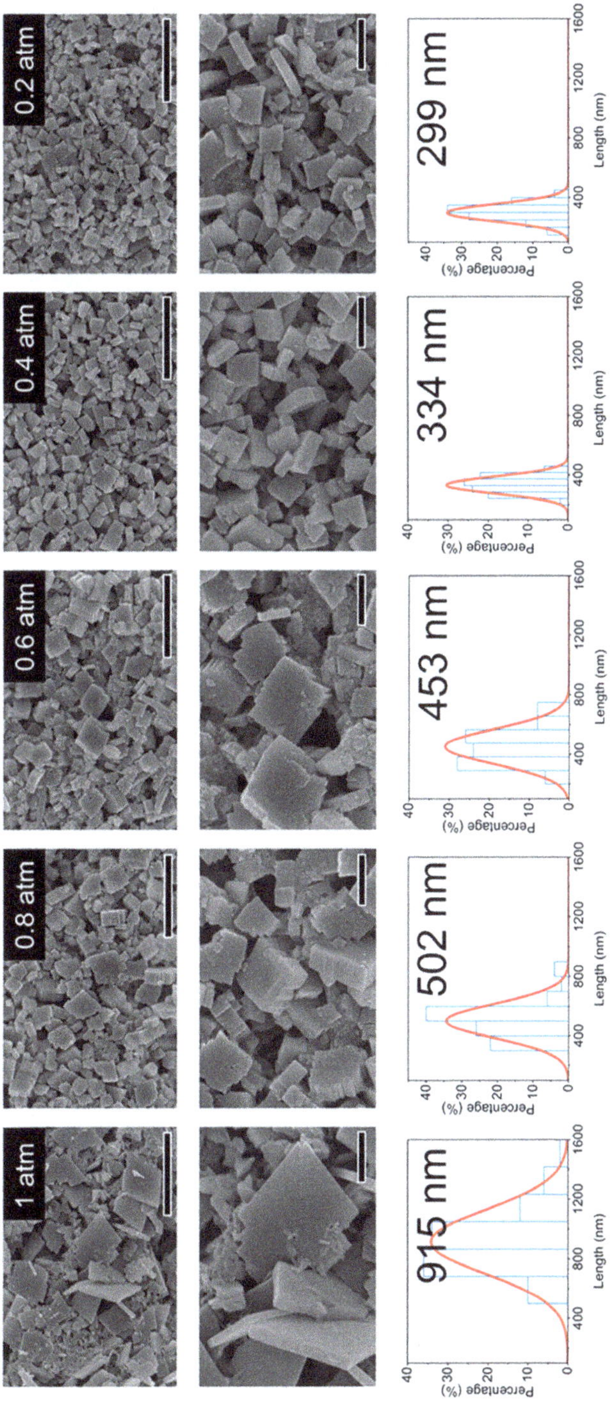

Figure 3.20: SEM images and the corresponding size distribution histograms of Cu-BDC synthesized under various pressures. Scale bars in SEM images: 2 μm (top panel) and 500 nm (middle panel) [44].

Recently, a CVD method was also used to fabricate uniform and mirror-like MOF thin films (Figure 3.21b–c) even in lift-off patterns or on fragile features [133]. To achieve that, they firstly vaporized 2-methylimidazole (HmIM) and then transported the vapor to the surface of ZnO-coated substrate, where the transformation of ZnO to ZIF-8 occurred.

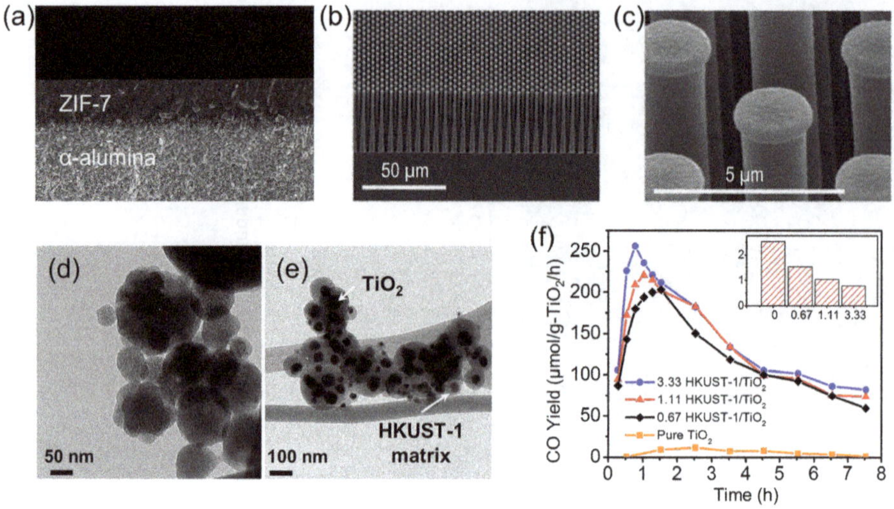

Figure 3.21: (a) SEM images of cross-sectional views of ZIF-7 membranes deposited on the α-alumina substrate [105]; (b) and (c) SEM images showing the ZIF-8-coated silicon pillar array [133]; TEM images of the as-synthesized HKUST-1 (d) and HKUST-1/TiO_2 (e); and (f) CO_2 photoreduction analysis of TiO_2 and HKUST-1/TiO_2 composites [41].

3.3.2.3 MOF-based composites

In addition, the aerosol process also enables the facile design of MOF-based materials without harsh chemicals, like hydrofluoric acid, which is commonly used in wet-chemistry methods [87]. Typical MOF-based composite materials prepared by the aerosol process include MOF/semiconductor [41–43, 80] and MOF/metal composites [37], which integrate the merits of both components and have been applied in many applications. For instance, HKUST-1/TiO_2 was fabricated in a fast manner by using the aerosol process (Figure 3.21d–e) [41]. The as-prepared HKUST-1/TiO_2 composite exhibits significantly improved photocatalytic efficiency toward CO_2 reduction (Figure 3.21f), primarily due to the enhanced molecule adsorption. Later, the efficiency of the composite was further boosted with the incorporation of a third semiconductor (i.e., Cu_2O) to facilitate the charge transfer [42].

3.3.2.4 Data-driven parameter optimization and perspectives

Despite all the advantages of the aerosol process for the synthesis of MOFs and MOF-based composites, there are still many challenges. For instance, the micro-droplet-based aerosol process or the so-called spray route typically completes in several seconds, which is disadvantageous for the synthesis of high-nuclearity MOFs, such as UiO-66 and MIL-100(Fe). These MOFs require a much longer reaction time to form the second building unit. To address this issue, an updated spray process was developed by introducing a continuous flow reactor before the atomizer, where the secondary building units form before the spray process [35]. With this updated spray process, both the yields and porosities of the MOF products were improved. In addition, there are many factors to be considered in the aerosol route, including both process and precursor parameters (Figure 3.22a). With such a large variety of parameters, a clear and logical optimization methodology is needed to ensure the success of the synthesis. To this end, He and coworkers developed a data science-driven approach to explore the dominating factors for the synthesis of high-quality MOFs by using several advanced computational methodologies, such as multidimensional scaling (MDS), MaxMin algorithm, Euclidean distance (ED), genetic algorithm (GA), and univariate feature selection (i.e., F-test) (Figure 3.22b) [40]. In particular, four major parameters that control the properties of MOFs were selected, that is, furnace temperature, operating pressure, solvent composition, and linker/metal ratio. To determine the significance of each parameter and achieve general guidance with the most representative parameter sets, the MaxMin algorithm was coupled with the pairwise ED to generate the most diverse 100 parameter sets [110], as can be visualized in the MDS plot in Figure 3.22c, where dots represent the parameter sets and the distance indicates the similarity. Then the synthesis of MOF samples (ZIF-8 as the model MOF) was carried out for the 21 most diverse parameter sets followed by full characterization to find out the successful samples. These successful parameter sets were subsequently fed into GA (e.g., migration, crossover, and mutation operations) to reproduce the second-generation parameter sets (Figure 3.22b). The optimization process was completed once the surface area of the MOF sample reaches the highest reported value. As shown in Figure 3.22d, the improved BET surface area and micropore area are observed for the second-generation MOF samples, demonstrating the effectiveness of the proposed data science approach to improve the aerosol synthesis of high-quality MOF products. The following F-test analysis further indicated that the furnace temperature plays the most dominant role among the four major parameters, which can be used to guide the future MOF fabrication via the aerosol route [40].

To further advance the fundamental understanding of the MOF formation via the aerosol route, future efforts should also be directed toward the elementary chemical steps from precursors to the final reticular structure with the aid of both experimental and simulation approaches, such as *operando* synchrotron scattering characterization, ab initio simulation, and machine learning [16, 40]. It is perceived that, with this

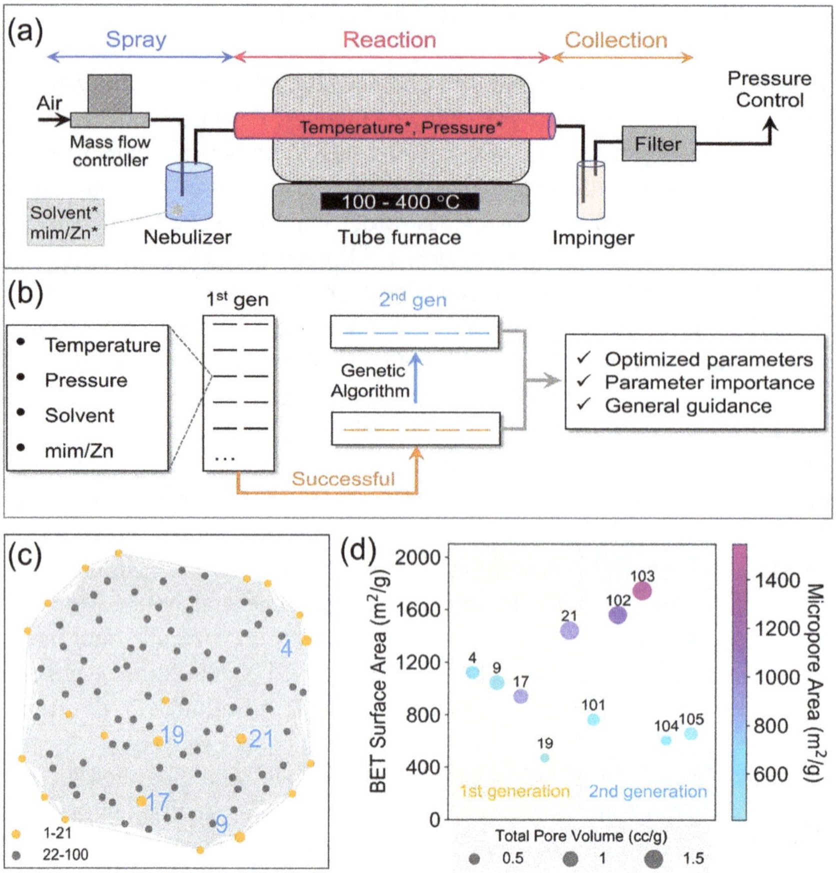

Figure 3.22: (a) Experimental setup of the spray system. The asterisk in the illustration denotes the synthesis parameters. (b) The logical flow of the data-driven parameter optimization approach. (c) MDS plot of the most distinct 100 experimental parameter sets, and (d) the BET surface area and micropore area of first- and second-generation MOF (ZIF-8 in this work) samples [40].

improved understanding of MOF formation in microdroplets, more hierarchical MOF structures would be achieved by using the aerosol processes in the near future.

3.4 Summary

This chapter summarizes the representative aerosol methodologies for the synthesis of advanced functional materials. Aerosol processing of materials via three major aerosol approaches, that is, gas-to-solid conversion, liquid-to-solid conversion, and aerosol-assisted self-assembly, is introduced. Strategies to control the size, size distribution, morphology, structure, composition, and functionality of the materials in those aerosol routes are discussed. Representative materials synthesized via the aerosol routes, such as nanoparticles, nanotubes, thin films, and hierarchical nanostructures, are described. Further, the aerosol routes are also promising for making emerging materials, such as CGO nanoballs and MOF-based materials. The applications of the aerosol-processed materials in the sectors of energy, the environment, and human health are also briefly discussed. Aerosol science and technology is ideally suited to address the challenges and issues in these areas due to its versatility and interdisciplinary nature.

Nomenclature

A,B	Constants
b	Boundary width of the grain
C	Droplet heat capacity
C_1, C_2	Concentrations of components 1 and 2, respectively
C_d	Concentration of the precursor (in droplet)
D	Fractal dimension
D_b	Diffusion coefficient of the grain boundary
D_g	Gas-phase diffusion coefficient
D_v	Diffusion coefficient of solvent vapor
d_p	Product particle diameter
d_d	Droplet diameter
$d_{d,i}$	Droplet initial diameter
F	Capillary force
k	Boltzmann constant
K_g	Thermal conductivity of the surrounding gas
K_l	Liquid electrical conductivity
m	Mass of the vapor molecule
m_d	Mass of a droplet
M_1, M_2	Molecular weights of components 1 and 2, respectively
M_g	Molecular weight of solvent
M_p	Molecular weight of the product particles
N_A	Avogadro constant
n_s, n_g	Vapor concentrations at droplet surface and in surrounding gas, respectively
P_d	Droplet pressure
p_v	Partial water vapor pressure
p_s	Saturated vapor pressure
Q	Liquid flow rate
r_d	Radius of droplet
r_p	Radius of particle
r^*	Electrical relaxation length
T_d	Droplet temperature
T_g	Surrounding gas temperature
$T(l)$	Temperature profile in a TiO_2 column as a function of its length l
T_m	Mixture temperature
u	Flow velocity
z	Spatial variable

Greek symbols

α	Solvent evaporation rate
γ	Surface tension
δ	Force scaling exponent
ε_0	Electrical permittivity of the vacuum
κ	Dielectric constant
λ	Latent heat of solvent evaporation
ρ_1, ρ_2	Densities of components 1 and 2, respectively
ρ_l	Liquid density
ρ_p	Density of the product particles
τ_{coll}	Particle–particle collision time
τ_{res}	Residence time
τ_{rxn}	Characteristic time for the reaction of the precursor
τ_{sin}	Sintering time
τ_{sl}	Characteristic time of solute diffusion
τ_{sv}	Characteristic time of solvent evaporation
Ω	Atomic volume

Subscripts

0	Vacuum
1, 2	Components 1 and 2, respectively
A	Avogadro
b	boundary
coll	collision
d	droplet
g	gas
i	initial
l	liquid
m	mixture
p	particle
res	residence
rxn	reaction
s	surface or saturated
sin	sintering
sl	solute diffusion
sv	solvent evaporation
v	vapor

References

[1] Amirav, L., Amirav, A., Lifshitz, E. (2005). A spray-based method for the production of semiconductor nanocrystals. *Journal of Physical Chemistry B* 109(20): 9857–9860.

[2] An, W.J., Jiang, D.D., Matthews, J.R., Borrelli, N.F., Biswas, P. (2011). Thermal conduction effects impacting morphology during synthesis of columnar nanostructured TiO_2 thin films. *Journal of Materials Chemistry* 21(22): 7913–7921.

[3] An, W.J., Thimsen, E., Biswas, P. (2010). Aerosol-chemical vapor deposition method for synthesis of nanostructured metal oxide thin films with controlled morphology. The *Journal of Physical Chemistry Letters* 1(1): 249–253.

[4] An, W.J., Wang, W.N., Ramalingam, B., Mukherjee, S., Daubayev, B., Gangopadhyay, S., Biswas, P. (2012). Enhanced water photolysis with Pt metal nanoparticles on single crystal TiO_2 surfaces. *Langmuir* 28(19): 7528–7534.

[5] Azam, M.A., Manaf, N.S.A., Talib, E., Bistamam, M.S.A. (2013). Aligned carbon nanotube from catalytic chemical vapor deposition technique for energy storage device: A review. *Ionics* 19 (11): 1455–1476.

[6] Balgis, R., Ogi, T., Arif, A.F., Anilkumar, G.M., Mori, T., Okuyama, K. (2015). Morphology control of hierarchical porous carbon particles from phenolic resin and polystyrene latex template via aerosol process. *Carbon* 84:281–289.

[7] Balgis, R., Ogi, T., Wang, W.N., Anilkumar, G.M., Sago, S., Okuyama, K. (2014). Aerosol synthesis of self-organized nanostructured hollow and porous carbon particles using a dual polymer system. *Langmuir* 30(38): 11257–11262.

[8] Barreiro, A., Selbmann, D., Pichler, T., Biedermann, K., Gemming, T., Rümmeli, M.H., Schwalke, U., Büchner, B. (2006). On the effects of solution and reaction parameters for the aerosol-assisted CVD growth of long carbon nanotubes. *Applied Physics A* 82(4): 719–725.

[9] Biswas, P., An, W.J., Wang, W.N. (2012). Nature's Nanostructures. Barnard, AS, Guo, H, eds, 443–476, Pan Stanford Publishing, Singapore.

[10] Brinker, C.J., Lu, Y.F., Sellinger, A., Fan, H.Y. (1999). Evaporation-induced self-assembly: Nanostructures made easy. *Advanced Materials* (Deerfield Beach, Fla.) 11(7): 579–585.

[11] Cao, X., Yin, Z., Zhang, H. (2014). Three-dimensional graphene materials: Preparation, structures and application in supercapacitors. *Energy & Environmental Science* 7(6): 1850–1865.

[12] Carné-Sánchez, A., Imaz, I., Cano-Sarabia, M., Maspoch, D. (2013). A spray-drying strategy for synthesis of nanoscale metal–organic frameworks and their assembly into hollow superstructures. *Nature Chemistry* 5(3): 203–211.

[13] Carné-Sánchez, A., Stylianou, K.C., Carbonell, C., Naderi, M., Imaz, I., Maspoch, D. (2015). Protecting Metal–Organic Framework Crystals from Hydrolytic Degradation by Spray-Dry Encapsulating Them into Polystyrene Microspheres. *Advanced Materials* (Deerfield Beach, Fla.) 27(5): 869–873.

[14] Chen, D.R., Pui, D.Y.H. (1997). Experimental investigation of scaling laws for electrospraying: Dielectric constant effect. *Aerosol Science and Technology* 27(3): 367–380.

[15] Chen, D.R., Pui, D.Y.H., Kaufman, S.L. (1995). Electrospraying of conducting liquids for monodisperse aerosol generation in the 4 nm to 1.8 μm diameter range. *Journal of Aerosol Science* 26(6): 963–977.

[16] Chen, J.P., Zhu, Z., Wang, W.N. (2021). Towards addressing environmental challenges: Rational design of metal-organic frameworks-based photocatalysts via a microdroplet approach. *Journal of Physics: Energy* 3:032005.

[17] Chen, Y.T., Guo, F., Jachak, A., Kim, S.P., Datta, D., Liu, J.Y., Kulaots, I., Vaslet, C., Jang, H.D., Huang, J.X., Kane, A., Shenoy, V.B., Hurt, R.H. (2012). Aerosol Synthesis of Cargo-Filled Graphene Nanosacks. 12(4): 1996–2002.

[18] Chen, Y.T., Guo, F., Qiu, Y., Hu, H.R., Kulaots, I., Walsh, E., Hurt, R.H. (2013). Encapsulation of Particle Ensembles in Graphene Nanosacks as a New Route to Multifunctional Materials. *ACS Nano* 7(5): 3744–3753.

[19] Choi, S.H., Jung, K.Y., Kang, Y.C. (2015). Amorphous GeO_x-coated reduced graphene oxide balls with sandwich structure for long-life lithium-ion batteries. *ACS Applied Materials & Interfaces* 7(25): 13952–13959.

[20] Choi, S.H., Kang, Y.C. (2014a). Crumpled Graphene–Molybdenum Oxide Composite Powders: Preparation and Application in Lithium-Ion Batteries. *Chemsuschem* 7(2): 523–528.

[21] Choi, S.H., Kang, Y.C. (2014b). Fe_3O_4-decorated hollow graphene balls prepared by spray pyrolysis process for ultrafast and long cycle-life lithium ion batteries. *Carbon* 79:58–66.

[22] Choi, S.H., Kang, Y.C. (2014c). Uniform Decoration of Vanadium Oxide Nanocrystals on Reduced Graphene-Oxide Balls by an Aerosol Process for Lithium-Ion Battery Cathode Material. *Chemistry – A European Journal* 20(21): 6294–6299.

[23] Choi, S.H., Ko, Y.N., Lee, J.-K., Kang, Y.C. (2014). Rapid continuous synthesis of spherical reduced graphene ball-nickel oxide composite for lithium ion batteries. *Sci Rep-UK* 4:5786.

[24] Cloupeau, M., Prunetfoch, B. (1994). Electrohydrodynamic spraying functioning modes – a critical-review. *Journal of Aerosol Science* 25(6): 1021–1036.

[25] Delamora, J.F., Loscertales, I.G. (1994). The current emitted by highly conducting Taylor cones. *Journal of Fluid Mechanics* 260:155–184.

[26] Dhakshinamoorthy, A., Asiri, A.M., Garcia, H. (2016). Metal-organic framework (MOF) compounds: Photocatalysts for redox reactions and solar fuel production. *Angewandte Chemie International Edition* 55(18): 5414–5445.

[27] Didenko, Y.T., Suslick, K.S. (2005). Chemical aerosol flow synthesis of semiconductor nanoparticles. *Journal of the American Chemical Society* 127(35): 12196–12197.

[28] Duch, M.C., Budinger, G.R.S., Liang, Y.T., Soberanes, S., Urich, D., Chiarella, S.E., Campochiaro, L.A., Gonzalez, A., Chandel, N.S., Hersam, M.C., Mutlu, G.M. (2011). Minimizing Oxidation and Stable Nanoscale Dispersion Improves the Biocompatibility of Graphene in the Lung. *Nano Letters* 11(12): 5201–5207.

[29] El-Shall, M.S., Li, S., Turkki, T., Graiver, D., Pernisz, U.C., Baraton, M.I. (1995). Synthesis and photoluminescence of weblike agglomeration of silica nanoparticles. *Journal of Physical Chemistry* 99(51): 17805–17809.

[30] Fisenko, S.P., Wang, W.N., Lenggoro, I.W., Okyuama, K. (2006). Evaporative cooling of micron-sized droplets in a low-pressure aerosol reactor. *Chemical Engineering Science* 61 (18): 6029–6034.

[31] Friedlander, S.K. (2000). Smoke, Dust, and Haze: Fundamental of Aerosol Dynamics, Oxford University Press, New York.

[32] Furukawa, H., Cordova, K.E., O'Keeffe, M., Yaghi, O.M. (2013). The Chemistry and Applications of Metal-Organic Frameworks. *Science* 341(6149): 1230444.

[33] Ganan-Calvo, A.M., Davila, J., Barrero, A. (1997). Current and droplet size in the electrospraying of liquids. Scaling laws. *Journal of Aerosol Science* 28(2): 249–275.

[34] Garcia Marquez, A., Horcajada, P., Grosso, D., Ferey, G., Serre, C., Sanchez, C., Boissiere, C. (2013). Green scalable aerosol synthesis of porous metal-organic frameworks. *Chemical Communications* 49(37): 3848–3850.

[35] Garzon-Tovar, L., Cano-Sarabia, M., Carne-Sanchez, A., Carbonell, C., Imaz, I., Maspoch, D. (2016). A spray-drying continuous-flow method for simultaneous synthesis and shaping of microspherical high nuclearity MOF beads. *Reaction Chemistry & Engineering* 1(5): 533–539.

[36] Garzón-Tovar, L., Rodríguez-Hermida, S., Imaz, I., Maspoch, D. (2017). Spray Drying for Making Covalent Chemistry: Postsynthetic Modification of Metal–Organic Frameworks. *Journal of the American Chemical Society* 139(2): 897–903.

[37] Gholampour, N., Chaemchuen, S., Hu, Z.-Y., Mousavi, B., Van Tendeloo, G., Verpoort, F. (2017). Simultaneous creation of metal nanoparticles in metal organic frameworks via spray drying technique. *Chemical Engineering Journal* 322:702–709.
[38] Haddad, K., Abokifa, A., Kavadiya, S., Chadha, T.S., Shetty, P., Wang, Y., Fortner, J., Biswas, P. (2016). Growth of single crystal, oriented SnO_2 nanocolumn arrays by aerosol chemical vapour deposition. *CrystEngComm* 18(39): 7544–7553.
[39] He, X., Chen, D.R., Wang, W.N. (2020). Bimetallic metal-organic frameworks (MOFs) synthesized using the spray method for tunable CO2 adsorption. *Chemical Engineering Journal* 382:122825.
[40] He, X., Chen, J.P., Albin, S., Zhu, Z., Wang, W.N. (2021). Data-driven parameter optimization for the synthesis of high-quality zeolitic imidazolate frameworks via a microdroplet route. *Advanced Powder Technology*: The International Journal of the Society of Powder Technology, Japan 32(1): 266–271.
[41] He, X., Gan, Z., Fisenko, S., Wang, D., El-Kaderi, H.M., Wang, W.-N. (2017). Rapid Formation of Metal–Organic Frameworks (MOFs) Based Nanocomposites in Microdroplets and Their Applications for CO_2 Photoreduction. *ACS Applied Materials & Interfaces* 9(11): 9688–9698.
[42] He, X., Wang, W.-N. (2018). MOF-based ternary nanocomposites for better CO2 photoreduction: Roles of heterojunctions and coordinatively unsaturated metal sites. *Journal of Materials Chemistry A* 6(3): 932–940.
[43] He, X., Wang, W.-N. (2019a). Rational Design of Efficient Semiconductor-based Photocatalysts via Microdroplets: A Review. *Kona Powder Part J* 36:201–214.
[44] He, X., Wang, W.N. (2019b). Pressure-regulated synthesis of Cu(TPA). (DMF) in microdroplets for selective CO2 adsorption. *Dalton Transactions* 48(3): 1006–1016.
[45] He, X., Wang, W.N. (2019c). Synthesis of Cu-Trimesic Acid/Cu-1,4-Benzenedioic Acid via Microdroplets: Role of Component Compositions. *Crystal Growth & Design* 19(2): 1095–1102.
[46] He, Y.B., Chen, F.L., Li, B., Qian, G.D., Zhou, W., Chen, B.L. (2018). Porous metal-organic frameworks for fuel storage. *Coordination Chemistry Reviews* 373:167–198.
[47] Herm, Z.R., Bloch, E.D., Long, J.R. (2014). Hydrocarbon Separations in Metal-Organic Frameworks. *Chemistry of Materials*: A Publication of the American Chemical Society 26(1): 323–338.
[48] Hidayat, D., Joni, I.M., Setianto, P.C., Okuyama, K. 2013. Synthesis of Lithium Cobalt Oxide using Low-Pressure Spray Pyrolysis. *Padjadjaran International Physics Symposium 2013 (Pips-2013): Contribution of Physics on Environmental and Energy Conservations* 1554: 87–89.
[49] Hidayat, D., Widiyastuti, W., Ogi, T., Okuyama, K. (2010). Droplet Generation and Nanoparticle Formation in Low-Pressure Spray Pyrolysis. *Aerosol Science and Technology* 44 (8): 692–705.
[50] Hinds, W.C. (1999). Aerosol Technology: Properties, Behavior, and Measurement of Airborne Particles, John Wiley & Sons, Inc., New York.
[51] Hong, D.-Y., Hwang, Y.K., Serre, C., Férey, G., Chang, J.-S. (2009). Porous Chromium Terephthalate MIL-101 with Coordinatively Unsaturated Sites: Surface Functionalization, Encapsulation, Sorption and Catalysis. *Advanced Functional Materials* 19(10): 1537–1552.
[52] Horcajada, P., Gref, R., Baati, T., Allan, P.K., Maurin, G., Couvreur, P., Ferey, G., Morris, R.E., Serre, C. (2012). Metal-Organic Frameworks in Biomedicine. *Chemical Reviews* 112(2): 1232–1268.
[53] Hu, Z.C., Deibert, B.J., Li, J. (2014). Luminescent metal-organic frameworks for chemical sensing and explosive detection. *Chemical Society Reviews* 43(16): 5815–5840.
[54] Hummers, W.S. Jr, Offeman, R.E. (1958). Preparation of graphitic oxide. *Journal of the American Chemical Society* 80(6): 1339–1339.

[55] Huxford, R.C., Della Rocca, J., Lin, W. (2010). Metal–organic frameworks as potential drug carriers. *Current Opinion in Chemical Biology* 14(2): 262–268.

[56] Iskandar, F., Gradon, L., Okuyama, K. (2003). Control of the morphology of nanostructured particles prepared by the spray drying of a nanoparticle sol. *Journal of Colloid and Interface Science* 265(2): 296–303.

[57] Iskandar, F., Mikrajuddin,, Okuyama, K. (2001). In situ production of spherical silica particles containing self-organized mesopores. *Nano Letters* 1(5): 231–234.

[58] Iskandar, F., Mikrajuddin,, Okuyama, K. (2002). Controllability of pore size and porosity on self-organized porous silica particles. *Nano Letters* 2(4): 389–392.

[59] Ito, Y., Nohira, T. (2000). Non-conventional electrolytes for electrochemical applications. *Electrochimica Acta* 45(15–16): 2611–2622.

[60] Itoh, Y., Abdullah, M., Okuyama, K. (2004). Direct preparation of nonagglomerated indium tin oxide nanoparticles using various spray pyrolysis methods. *Journal of Materials Research* 19 (4): 1077–1086.

[61] Itoh, Y., Lenggoro, I.W., Okuyama, K., Madler, L., Pratsinis, S.E. (2003). Size tunable synthesis of highly crystalline BaTiO3 nanoparticles using salt-assisted spray pyrolysis. *Journal of Nanoparticle Research*: An Interdisciplinary Forum for Nanoscale Science and Technology 5(3–4): 191–198.

[62] Itoh, Y., Lenggoro, I.W., Pratsinis, S.E., Okuyama, K. (2002). Agglomerate-free $BaTiO_3$ particles by salt-assisted spray pyrolysis. *Journal of Materials Research* 17(12): 3222–3229.

[63] Itoh, Y., Okuyama, K. (2003). Preparation of agglomerate-free and highly crystalline ($Ba_{0.5}$, $Sr_{0.5}$)TiO_3 nanoparticles by salt-assisted spray pyrolysis. *Journal of the Ceramic Society of Japan* 111(11): 815–820.

[64] Jiang, J., Chen, D.R., Biswas, P. (2007). Synthesis of nanoparticles in a flame aerosol reactor with independent and strict control of their size, crystal phase and morphology. *Nanotechnology* 18(28): 285603.

[65] Jiang, Y., Liu, D., Cho, M., Lee, S.S., Zhang, F., Biswas, P., Fortner, J.D. (2016). In Situ Photocatalytic Synthesis of Ag Nanoparticles (nAg) by Crumpled Graphene Oxide Composite Membranes for Filtration and Disinfection Applications. *Environmental Science & Technology* 50(5): 2514–2521.

[66] Jiang, Y., Wang, W.-N., Biswas, P., Fortner, J.D. (2014). Facile Aerosol Synthesis and Characterization of Ternary Crumpled Graphene–TiO_2–Magnetite Nanocomposites for Advanced Water Treatment. *ACS Applied Materials & Interfaces* 6(14): 11766–11774.

[67] Jiang, Y., Wang, W.-N., Liu, D., Nie, Y., Li, W., Wu, J., Zhang, F., Biswas, P., Fortner, J.D. (2015). Engineered Crumpled Graphene Oxide Nanocomposite Membrane Assemblies for Advanced Water Treatment Processes. *Environmental Science & Technology* 49(11): 6846–6854.

[68] Jo, E.H., Choi, J.-H., Park, S.-R., Lee, C.M., Chang, H., Jang, H.D. (2016). Size and Structural Effect of Crumpled Graphene Balls on the Electrochemical Properties for Supercapacitor Application. *Electrochimica Acta* 222:58–63.

[69] Jo, E.H., Jang, H.D., Chang, H., Kim, S.K., Choi, J.-H., Lee, C.M. (2017). 3 D Network-Structured Crumpled Graphene/Carbon Nanotube/Polyaniline Composites for Supercapacitors. *Chemsuschem* 10(10): 2210–2217.

[70] Kang, H.S., Kang, Y.C., Koo, H.Y., Ju, S.H., Kim, D.Y., Hong, S.K., Sohn, J.R., Jung, K.Y., Park, S.B. (2006). Nano-sized ceria particles prepared by spray pyrolysis using polymeric precursor solution. *Mat Sci Eng B-Solid* 127(2–3): 99–104.

[71] Kang, Y.C., Park, S.B. (1996). Preparation of nanometre size oxide particles using filter expansion aerosol generator. *J Mater Sci* 31(9): 2409–2416.

[72] Kavadiya, S., Raliya, R., Schrock, M., Biswas, P. (2017). Crumpling of graphene oxide through evaporative confinement in nanodroplets produced by electrohydrodynamic aerosolization.

Journal of Nanoparticle Research: An Interdisciplinary Forum for Nanoscale Science and Technology 19(2): 43.
[73] Khoshaman, A.H., Bahreyni, B. (2012). Application of metal organic framework crystals for sensing of volatile organic gases. *Sensors and Actuators. B, Chemical* 162(1): 114–119.
[74] Kim, J.H., Germer, T.A., Mulholland, G.W., Ehrman, S.H. (2002). Size-monodisperse metal nanoparticles via hydrogen-free spray pyrolysis. *Advanced Materials* (Deerfield Beach, Fla.) 14(7): 518–521.
[75] Kodas, T.T., Hampden-Smith, M. (1999). Aerosol Processing of Materials, Wiley-VCH, New York.
[76] Kondo, M., Yoshitomi, T., Seki, K., Matsuzaka, H., Kitagawa, S. (1997). Three-dimensional framework with channeling cavities for small molecules: {[M-2(4,4 '-bpy)(3)(NO3)(4)]center dot xH(2)O}(n) (M = Co, Ni, Zn). *Angewandte Chemie International Edition* 36(16): 1725–1727.
[77] Koós, A.A., Dowling, M., Jurkschat, K., Crossley, A., Grobert, N. (2009). Effect of the experimental parameters on the structure of nitrogen-doped carbon nanotubes produced by aerosol chemical vapour deposition. *Carbon* 47(1): 30–37.
[78] Kortshagen, U.R., Sankaran, R.M., Pereira, R.N., Girshick, S.L., Wu, J.J., Aydil, E.S. (2016). Nonthermal Plasma Synthesis of Nanocrystals: Fundamental Principles. Materials, and Applications. *Chem Rev* 116(18): 11061–11127.
[79] Kreno, L.E., Leong, K., Farha, O.K., Allendorf, M., Van Duyne, R.P., Hupp, J.T. (2012). Metal-Organic Framework Materials as Chemical Sensors. *Chemical Reviews* 112(2): 1105–1125.
[80] Kubo, M., Moriyama, R., Shimada, M. (2019). Facile fabrication of HKUST-1 nanocomposites incorporating Fe3O4 and TiO2 nanoparticles by a spray-assisted synthetic process and their dye adsorption performances. *Microporous and Mesoporous Materials* 280:227–235.
[81] Kubo, M., Saito, T., Shimada, M. (2017). Evaluation of the parameters utilized for the aerosol-assisted synthesis of HKUST-1. *Microporous and Mesoporous Materials* 245:126–132.
[82] Kumar, M., Ando, Y. (2010). Chemical vapor deposition of carbon nanotubes: A review on growth mechanism and mass production. *Journal of Nanoscience and Nanotechnology* 10(6): 3739–3758.
[83] Lee, S.Y., Gradon, L., Janeczko, S., Iskandar, F., Okuyama, K. (2010). Formation of Highly Ordered Nanostructures by Drying Micrometer Colloidal Droplets. *ACS Nano* 4(8): 4717–4724.
[84] Lenggoro, I.W., Hata, T., Iskandar, F., Lunden, M.M., Okuyama, K. (2000). An experimental and modeling investigation of particle production by spray pyrolysis using a laminar flow aerosol reactor. *Journal of Materials Research* 15(3): 733–743.
[85] Lenggoro, I.W., Itoh, Y., Iida, N., Okuyama, K. (2003). Control of size and morphology in NiO particles prepared by a low-pressure spray pyrolysis. *Materials Research Bulletin* 38(14): 1819–1827.
[86] Li, J.R., Sculley, J., Zhou, H.C. (2012). Metal-Organic Frameworks for Separations. *Chemical Reviews* 112(2): 869–932.
[87] Li, R., Hu, J., Deng, M., Wang, H., Wang, X., Hu, Y., Jiang, H.-L., Jiang, J., Zhang, Q., Xie, Y., Xiong, Y. (2014). Integration of an Inorganic Semiconductor with a Metal–Organic Framework: A Platform for Enhanced Gaseous Photocatalytic Reactions. *Advanced Materials* (Deerfield Beach, Fla.) 26(28): 4783–4788.
[88] Li, S.Q., Ren, Y.H., Biswas, P., Tse, S.D. (2016). Flame aerosol synthesis of nanostructured materials and functional devices: Processing, modeling, and diagnostics. *Progress in Energy and Combustion* 55:1–59.
[89] Li, X., Yu, J.G., Jaroniec, M., Chen, X.B. (2019). Cocatalysts for Selective Photoreduction of CO_2 into Solar Fuels. *Chemical Reviews* 119(6): 3962–4179.

[90] Li, Y.-N., Wang, S., Zhou, Y., Bai, X.-J., Song, G.-S., Zhao, X.-Y., Wang, T.-Q., Qi, X., Zhang, X.-M., Fu, Y. (2017). Fabrication of Metal–Organic Framework and Infinite Coordination Polymer Nanosheets by the Spray Technique. *Langmuir* 33(4): 1060–1065.
[91] Liu, S., Wang, A., Li, Q., Wu, J., Chiou, K., Huang, J., Luo, J. (2018). Crumpled Graphene Balls Stabilized Dendrite-free Lithium Metal Anodes. *Joule* 2(1): 184–193.
[92] Lu, Y.F., Fan, H.Y., Stump, A., Ward, T.L., Rieker, T., Brinker, C.J. (1999). Aerosol-assisted self-assembly of mesostructured spherical nanoparticles. *Nature* 398(6724): 223–226.
[93] Luo, J., Jang, H.D., Huang, J. (2013). Effect of Sheet Morphology on the Scalability of Graphene-Based Ultracapacitors. *ACS Nano* 7(2): 1464–1471.
[94] Luo, J., Jang, H.D., Sun, T., Xiao, L., He, Z., Katsoulidis, A.P., Kanatzidis, M.G., Gibson, J.M., Huang, J. (2011). Compression and Aggregation-Resistant Particles of Crumpled Soft Sheets. *ACS Nano* 5(11): 8943–8949.
[95] Luo, J., Zhao, X., Wu, J., Jang, H.D., Kung, H.H., Huang, J. (2012). Crumpled Graphene-Encapsulated Si Nanoparticles for Lithium Ion Battery Anodes. *The Journal of Physical Chemistry Letters* 3(13): 1824–1829.
[96] Ma, X., Zachariah, M.R., Zangmeister, C.D. (2012). Crumpled Nanopaper from Graphene Oxide. *Nano Letters* 12(1): 486–489.
[97] Machado, B.F., Serp, P. (2012). Graphene-based materials for catalysis. *Catalysis Science & Technology* 2(1): 54–75.
[98] Mann, A.K.P., Skrabalak, S.E. (2011). Synthesis of Single-Crystalline Nanoplates by Spray Pyrolysis: A Metathesis Route to Bi_2WO_6. Chemistry of Materials: A Publication of the American *Chemical Society* 23(4): 1017–1022.
[99] Mann, A.K.P., Wicker, S., Skrabalak, S.E. (2012). Aerosol-Assisted Molten Salt Synthesis of $NaInS_2$ Nanoplates for Use as a New Photoanode Material. *Advanced Materials* (Deerfield Beach, Fla.) 24(46): 6186–6191.
[100] Mao, B.S., Wen, Z., Bo, Z., Chang, J., Huang, X., Chen, J. (2014). Hierarchical Nanohybrids with Porous CNT-Networks Decorated Crumpled Graphene Balls for Supercapacitors. *ACS Applied Materials & Interfaces* 6(12): 9881–9889.
[101] Mao, S., Lu, G., Chen, J. (2015). Three-dimensional graphene-based composites for energy applications. *Nanoscale* 7(16): 6924–6943.
[102] Maruyama, S., Kojima, R., Miyauchi, Y., Chiashi, S., Kohno, M. (2002). Low-temperature synthesis of high-purity single-walled carbon nanotubes from alcohol. *Chemical Physics Letters* 360(3): 229–234.
[103] Mattox, D.M. (2010). Handbook of Physical Vapor Deposition (PVD) Processing, Elsevier, Oxford, UK.
[104] Melgar, V.M.A., Ahn, H., Kim, J., Othman, M.R. (2015). Highly selective micro-porous ZIF-8 membranes prepared by rapid electrospray deposition. *J Ind Eng Chem* 21:575–579.
[105] Melgar, V.M.A., Kwon, H.T., Kim, J. (2014). Direct spraying approach for synthesis of ZIF-7 membranes by electrospray deposition. The *Journal of Membrane Science* 459:190–196.
[106] Messing, G.L., Zhang, S.C., Jayanthi, G.V. (1993). Ceramic Powder Synthesis by Spray-Pyrolysis. *Journal of the American Ceramic Society*. American Ceramic Society 76(11): 2707–2726.
[107] Meysami, S.S., Koós, A.A., Dillon, F., Grobert, N. (2013). Aerosol-assisted chemical vapour deposition synthesis of multi-wall carbon nanotubes: II. An analytical study. *Carbon* 58: 159–169.
[108] Moghadam, P.Z., Li, A., Wiggin, S.B., Tao, A., Maloney, A.G.P., Wood, P.A., Ward, S.C., Fairen-Jimenez, D. (2017). Development of a Cambridge Structural Database Subset: A Collection of Metal–Organic Frameworks for Past, Present, and Future. *Chemistry of Materials*: A Publication of the American Chemical Society 29(7): 2618–2625.

[109] Moore, T.L., Podilakrishna, R., Rao, A., Alexis, F. (2014). Systemic Administration of Polymer-Coated Nano-Graphene to Deliver Drugs to Glioblastoma. *Particle & Particle Systems Characterization* 31(8): 886–894.
[110] Moosavi, S.M., Chidambaram, A., Talirz, L., Haranczyk, M., Stylianou, K.C., Smit, B. (2019). Capturing chemical intuition in synthesis of metal-organic frameworks. *Nature Communications* 10:539.
[111] Nakaso, K., Okuyama, K., Shimada, M., Pratsinis, S.E. (2003). Effect of reaction temperature on CVD-made TiO_2 primary particle diameter. *Chemical Engineering Science* 58(15): 3327–3335.
[112] Nandiyanto, A.B.D., Ogi, T., Wang, W.N., Gradon, L., Okuyama, K. (2019). Template-assisted spray-drying method for the fabrication of porous particles with tunable structures. *Advanced Powder Technology*: The International Journal of the Society of Powder Technology, Japan 30(12): 2908–2924.
[113] Nandiyanto, A.B.D., Okuyama, K. (2011). Progress in developing spray-drying methods for the production of controlled morphology particles: From the nanometer to submicrometer size ranges. *Advanced Powder Technology*: The International Journal of the Society of Powder Technology, Japan 22(1): 1–19.
[114] Nandiyanto, A.B.D., Suhendi, A., Arutanti, O., Ogi, T., Okuyama, K. (2013). Influences of Surface Charge, Size, and Concentration of Colloidal Nanoparticles on Fabrication of Self-Organized Porous Silica in Film and Particle Forms. *Langmuir* 29(21): 6262–6270.
[115] Nandiyanto, A.B.D., Suhendi, A., Ogi, T., Umemoto, R., Okuyama, K. (2014). Size- and charge-controllable polystyrene spheres for templates in the preparation of porous silica particles with tunable internal hole configurations. *Chemical Engineering Journal* 256:421–430.
[116] Nazarian-Samani, M., Kim, H.-K., Park, S.-H., Youn, H.-C., Mhamane, D., Lee, S.-W., Kim, M.-S., Jeong, J.-H., Haghighat-Shishavan, S., Roh, K.-C., Kashani-Bozorg, S.F., Kim, K.-B. (2016). Three-dimensional graphene-based spheres and crumpled balls: Micro- and nano-structures, synthesis strategies, properties and applications. *RSC Advances* 6(56): 50941–50967.
[117] Nie, Y., Wang, W.-N., Jiang, Y., Fortner, J., Biswas, P. (2016). Crumpled reduced graphene oxide-amine-titanium dioxide nanocomposites for simultaneous carbon dioxide adsorption and photoreduction. *Catalysis Science & Technology* 6(16): 6187–6196.
[118] Novoselov, K.S., Geim, A.K., Morozov, S.V., Jiang, D., Zhang, Y., Dubonos, S.V., Grigorieva, I.V., Firsov, A.A. (2004). Electric field effect in atomically thin carbon films. *Science* 306 (5696): 666–669.
[119] Ogi, T., Hidayat, D., Iskandar, F., Purwanto, A., Okuyama, K. (2009a). Direct synthesis of highly crystalline transparent conducting oxide nanoparticles by low pressure spray pyrolysis. *Advanced Powder Technology*: The International Journal of the Society of Powder Technology, Japan 20(2): 203–209.
[120] Ogi, T., Iskandar, F., Itoh, Y., Okuyama, K. (2006). Characterization of dip-coated ITO films derived from nanoparticles synthesized by low-pressure spray pyrolysis. *Journal of Nanoparticle Research*: An Interdisciplinary Forum for Nanoscale Science and Technology 8 (3–4): 343–350.
[121] Ogi, T., Kaihatsu, Y., Iskandar, F., Tanabe, E., Okuyama, K. (2009b). Synthesis of nanocrystalline GaN from Ga_2O_3 nanoparticles derived from salt-assisted spray pyrolysis. *Advanced Powder Technology*: The International Journal of the Society of Powder Technology, Japan 20(1): 29–34.

[122] Okuyama, K., Kousaka, Y., Tohge, N., Yamamoto, S., Wu, J.J., Flagan, R.C., Seinfeld, J.H. (1986). Production of Ultrafine Metal-Oxide Aerosol-Particles by Thermal-Decomposition of Metal Alkoxide Vapors. *AIChE Journal*. American Institute of Chemical Engineers 32(12): 2010–2019.

[123] Okuyama, K., Lenggoro, I.W. (2003). Preparation of nanoparticles via spray route. Chemical Engineering Science 58(3–6): 537–547.

[124] Okuyama, K., Ushio, R., Kousaka, Y., Flagan, R.C., Seinfeld, J.H. (1990). Particle Generation in a Chemical Vapor-Deposition Process with Seed Particles. *AIChE Journal*. American Institute of Chemical Engineers 36(3): 409–419.

[125] Pierson, H.O. (1999). Handbook of Chemical Vapor Deposition: Principles, Technology, and Applications, Noyes Publications/William Andrew Publishing, LLC, New York.

[126] Pratsinis, S.E. (1998). Flame aerosol synthesis of ceramic powders. *Progress in Energy and Combustion* 24(3): 197–219.

[127] Reddy, A.L.M., Srivastava, A., Gowda, S.R., Gullapalli, H., Dubey, M., Ajayan, P.M. (2010). Synthesis of nitrogen-doped graphene films for lithium battery application. *ACS Nano* 4(11): 6337–6342.

[128] Robinson, J.T., Tabakman, S.M., Liang, Y., Wang, H., Sanchez Casalongue, H., Vinh, D., Dai, H. (2011). Ultrasmall reduced graphene oxide with high near-infrared absorbance for photothermal therapy. *Journal of the American Chemical Society* 133(17): 6825–6831.

[129] Rosell-Llompart, J., Delamora, J.F. (1994). Generation of monodisperse droplets 0.3 to 4 μM in diameter from electrified cone-jets of highly conducting and viscous-liquids. *Journal of Aerosol Science* 25(6): 1093–1119.

[130] Shah, K.A., Tali, B.A. (2016). Synthesis of carbon nanotubes by catalytic chemical vapour deposition: A review on carbon sources, catalysts and substrates. *Materials Science in Semiconductor Processing* 41:67–82.

[131] Shimada, M., Wang, W.N., Okuyama, K. (2010). Synthesis of Gallium Nitride Nanoparticles by Microwave Plasma-Enhanced CVD. Chemical Vapor Deposition 16(4–6): 151–156.

[132] Sinnott, S.B., Andrews, R., Qian, D., Rao, A.M., Mao, Z., Dickey, E.C., Derbyshire, F. (1999). Model of carbon nanotube growth through chemical vapor deposition. *Chemical Physics Letters* 315(1): 25–30.

[133] Stassen, I., Styles, M., Grenci, G., Gorp, H.V., Vanderlinden, W., Feyter, S.D., Falcaro, P., Vos, D.D., Vereecken, P., Ameloot, R. (2016). Chemical vapour deposition of zeolitic imidazolate framework thin films. *Nature Materials* 15(3): 304–310.

[134] Strobel, R., Pratsinis, S.E. (2007). Flame aerosol synthesis of smart nanostructured materials. *Journal of Materials Chemistry* 17(45): 4743–4756.

[135] Suhendi, A., Nandiyanto, A.B.D., Munir, M.M., Ogi, T., Gradon, L., Okuyama, K. (2013). Self-Assembly of Colloidal Nanoparticles Inside Charged Droplets during Spray-Drying in the Fabrication of Nanostructured Particles. *Langmuir* 29(43): 13152–13161.

[136] Sumida, K., Rogow, D.L., Mason, J.A., McDonald, T.M., Bloch, E.D., Herm, Z.R., Bae, T.-H., Long, J.R. (2012). Carbon Dioxide Capture in Metal–Organic Frameworks. *Chemical Reviews* 112(2): 724–781.

[137] Tang, H., Yang, C., Lin, Z., Yang, Q., Kang, F., Wong, C.P. (2015). Electrospray-deposition of graphene electrodes: A simple technique to build high-performance supercapacitors. *Nanoscale* 7(20): 9133–9139.

[138] Tarascon, J.M., Armand, M. (2001). Issues and challenges facing rechargeable lithium batteries. *Nature* 414:359.

[139] Terashi, Y., Purwanto, A., Wang, W.N., Iskandar, F., Okuyama, K. (2008). Role of urea addition in the preparation of tetragonal $BaTiO_3$ nanoparticles using flame-assisted spray pyrolysis. *Journal of the European Ceramic Society* 28(13): 2573–2580.

[140] Thimsen, E., Biswas, P. (2007). Nanostructured photoactive films synthesized by a flame aerosol reactor. *AIChE Journal*. American Institute of Chemical Engineers 53(7): 1727–1735.
[141] Thimsen, E., Rastgar, N., Biswas, P. (2007). Rapid synthesis of nanostructured metal-oxide films for solar energy applications by a flame aerosol reactor (FLAR). Proceedings of SPIE 6650:66500G.
[142] Thimsen, E., Rastgar, N., Biswas, P. (2008). Nanostructured TiO_2 films with controlled morphology synthesized in a single step process: Performance of dye-sensitized solar cells and photo watersplitting. *Journal of Physical Chemistry C* 112(11): 4134–4140.
[143] Tonelli, F.M.P., Goulart, V.A.M., Gomes, K.N., Ladeira, M.S., Santos, A.K., Lorençon, E., Ladeira, L.O., Resende, R.R. (2015). Graphene-based nanomaterials: Biological and medical applications and toxicity. *Nanomedicine-UK* 10(15): 2423–2450.
[144] Wang, L., Lu, C., Zhang, B., Zhao, B., Wu, F., Guan, S. (2014a). Fabrication and characterization of flexible silk fibroin films reinforced with graphene oxide for biomedical applications. *RSC Advances* 4(76): 40312–40320.
[145] Wang, W.-N., Jiang, Y., Biswas, P. (2012a). Evaporation-Induced Crumpling of Graphene Oxide Nanosheets in Aerosolized Droplets: Confinement Force Relationship. *The Journal of Physical Chemistry Letters* 3(21): 3228–3233.
[146] Wang, W.-N., Jiang, Y., Fortner, J.D., Biswas, P. (2014b). Nanostructured Graphene-Titanium Dioxide Composites Synthesized by a Single-Step Aerosol Process for Photoreduction of Carbon Dioxide. *Environmental Engineering Science* 31(7): 428–434.
[147] Wang, W.-N., Widiyastuti, W., Lenggoro, I.W., Kim, T.O., Okuyama, K. (2007a). Photoluminescence optimization of luminescent nanocomposites fabricated by spray pyrolysis of a colloid-solution precursor. *Journal of the Electrochemical Society* 154(4): J121–J128.
[148] Wang, W.N., An, W.J., Ramalingam, B., Mukherjee, S., Niedzwiedzki, D.M., Gangopadhyay, S., Biswas, P. (2012b). Size and Structure Matter: Enhanced CO_2 Photoreduction Efficiency by Size-Resolved Ultrafine Pt Nanoparticles on TiO_2 Single Crystals. *Journal of the American Chemical Society* 134(27): 11276–11281.
[149] Wang, W.N., Itoh, Y., Lenggoro, I.W., Okuyama, K. (2004). Nickel and nickel oxide nanoparticles prepared from nickel nitrate hexahydrate by a low pressure spray pyrolysis. *Mat Sci Eng B-Solid* 111(1): 69–76.
[150] Wang, W.N., Kim, S.G., Lenggoro, I.W., Okuyama, K. (2007b). Polymer-assisted annealing of spray-pyrolyzed powders for formation of luminescent particles with submicrometer and nanometer sizes. *Journal of the American Ceramic Society*. American Ceramic Society 90(2): 425–432.
[151] Wang, W.N., Lenggoro, I.W., Okuyama, K. (2005a). Dispersion and aggregation of nanoparticles derived from colloidal droplets under low-pressure conditions. *Journal of Colloid and Interface Sciencei* 288(2): 423–431.
[152] Wang, W.N., Lenggoro, I.W., Okuyama, K. (2011). Preparation of Nanoparticles by Spray Routes, in Encyclopedia of Nanoscience and Nanotechnology. Nalwa, H.S., ed, 435–458, American Scientific Publishers, Stevenson Ranch, California.
[153] Wang, W.N., Lenggoro, I.W., Okuyama, K., Terashi, Y., Wang, Y.C. (2006). Effects of ethanol addition and Ba/Ti ratios on preparation of barium titanate nanocrystals via a spray pyrolysis method. *Journal of the American Ceramic Society*. American Ceramic Society 89(3): 888–893.
[154] Wang, W.N., Lenggoro, I.W., Terashi, Y., Wang, Y.C., Okuyama, K. (2005b). Direct synthesis of barium titanate nanoparticles via a low pressure spray pyrolysis method. *Journal of Materials Research* 20(10): 2873–2882.
[155] Wang, W.N., Soulis, J., Yang, Y.J., Biswas, P. (2014c). Comparison of CO_2 Photoreduction Systems: A Review. *Aerosol and Air Quality Research* 14(2): 533–549.

[156] Wang, Y., Liu, P., Fang, J.X., Wang, W.N., Biswas, P. (2015a). Kinetics of sub-2 nm TiO_2 particle formation in an aerosol reactor during thermal decomposition of titanium tetraisopropoxide. *Journal of Nanoparticle Research*: An Interdisciplinary Forum for Nanoscale Science and Technology 17(3): 147.
[157] Wang, Y., Shao, Y., Matson, D.W., Li, J., Lin, Y. (2010). Nitrogen-Doped Graphene and Its Application in Electrochemical Biosensing. *ACS Nano* 4(4): 1790–1798.
[158] Wang, Z., Ananias, D., Carné-Sánchez, A., Brites, C.D.S., Imaz, I., Maspoch, D., Rocha, J., Carlos, L.D. (2015b). Lanthanide–organic framework nanothermometers prepared by spray-drying. *Advanced Functional Materials* 25(19): 2824–2830.
[159] Wang, Z., Zhou, L., Lou, X.W. (2012c). Metal Oxide Hollow Nanostructures for Lithium-ion Batteries. *Advanced Materials* (Deerfield Beach, Fla.) 24(14): 1903–1911.
[160] Widiyastuti, W., Wang, W.N., Purwanto, A., Lenggoro, I.W., Okuyama, K. (2007). A pulse combustion-spray pyrolysis process for the preparation of nano- and submicrometer-sized oxide particles. *Journal of the American Ceramic Society*. American Ceramic Society 90(12): 3779–3785.
[161] Worathanakul, P., Jiang, J.K., Biswas, P., Kongkachuichay, P. (2008). Quench-Ring Assisted Flame Synthesis of SiO_2-TiO_2 Nanostructured Composite. *Journal for Nanoscience and Nanotechnology* 8(12): 6253–6259.
[162] Xia, B., Lenggoro, I.W., Okuyama, K. (2001a). Novel Route to Nanoparticle Synthesis by Salt-Assisted Aerosol Decomposition. *Advanced Materials* (Deerfield Beach, Fla.) 13(20): 1579–1582.
[163] Xia, B., Lenggoro, I.W., Okuyama, K. (2001b). Synthesis of CeO_2 nanoparticles by salt-assisted ultrasonic aerosol decomposition. *Journal of Materials Chemistry* 11(12): 2925–2927.
[164] Xia, B., Lenggoro, I.W., Okuyama, K. (2002a). Nanoparticle separation in salted droplet microreactors. *Chemistry of Materials*: A Publication of the American Chemical Society 14(6): 2623–2627.
[165] Xia, B., Lenggoro, I.W., Okuyama, K. (2002b). Synthesis and photoluminescence of spherical ZnS: Mn^{2+} particles. *Chemistry of Materials*: A Publication of the American Chemical Society 14(12): 4969–4974.
[166] Xiang, R., Einarsson, E., Okawa, J., Miyauchi, Y., Maruyama, S. (2009). Acetylene-Accelerated Alcohol Catalytic Chemical Vapor Deposition Growth of Vertically Aligned Single-Walled Carbon Nanotubes. *The Journal of Physical Chemistry C* 113(18): 7511–7515.
[167] Xin, S., Guo, Y.-G., Wan, L.-J. (2012). Nanocarbon Networks for Advanced Rechargeable Lithium Batteries. *Accounts of Chemical Research* 45(10): 1759–1769.
[168] Yaghi, O.M., O'Keeffe, M., Ockwig, N.W., Chae, H.K., Eddaoudi, M., Kim, J. (2003). Reticular synthesis and the design of new materials. *Nature* 423(6941): 705–714.
[169] Yan, L., Zhao, F., Li, S., Hu, Z., Zhao, Y. (2011). Low-toxic and safe nanomaterials by surface-chemical design. Carbon Nanotubes, Fullerenes, Metallofullerenes, and Graphenes. *Nanoscale* 3(2): 362–382.
[170] Yang, K., Li, Y., Tan, X., Peng, R., Liu, Z. (2013). Behavior and Toxicity of Graphene and Its Functionalized Derivatives in Biological Systems. *Small* 9(9–10): 1492–1503.
[171] Yoon, J.W., Seo, Y.-K., Hwang, Y.K., Chang, J.-S., Leclerc, H., Wuttke, S., Bazin, P., Vimont, A., Daturi, M., Bloch, E., Llewellyn, P.L., Serre, C., Horcajada, P., Grenèche, J.-M., Rodrigues, A.E., Férey, G. (2010). Controlled Reducibility of a Metal–Organic Framework with Coordinatively Unsaturated Sites for Preferential Gas Sorption. *Angewandte Chemie International Edition* 49(34): 5949–5952.
[172] Zhu, L., Liu, X.-Q., Jiang, H.-L., Sun, L.-B. (2017). Metal–Organic Frameworks for Heterogeneous Basic Catalysis. *Chemical Reviews* 117(12): 8129–8176.

[173] Zhu, Z., Chen, J., Wang, W.-N. (2021). Artificial Photosynthesis by 3D Graphene-based Composite Photocatalysts, in Graphene-based 3D Macrostructures for Clean Energy and Environmental Applications. Balasubramanian, R., Chowdhury, S., eds, 396–431, The Royal Society of Chemistry, Cambridge, UK.
[174] Zodrow, K., Brunet, L., Mahendra, S., Li, D., Zhang, A., Li, Q., Alvarez, P.J.J. (2009). Polysulfone ultrafiltration membranes impregnated with silver nanoparticles show improved biofouling resistance and virus removal. *Water Research* 43(3): 715–723.

Jiayu Li and Pratim Biswas

Chapter 4 Calibration and applications of low-cost particle sensors: a review of recent advances

Abstract: Particulate matter (PM) has been monitored routinely due to its negative effects on human health and atmospheric visibility. Standard gravimetric measurements and current commercial instruments for field measurements are still expensive and laborious. The high cost of conventional instruments typically limits the number of monitoring sites, which in turn undermines the accuracy of real-time mapping with insufficient spatial resolution. The new trends of PM concentration measurement are personalized portable devices for individual customers and the networking of large quantity sensors to meet the demand of Big Data. Therefore, low-cost PM sensors have been studied extensively due to their price advantage and compact size. These sensors have been considered as a good supplement to current monitoring sites for high spatial-temporal PM mapping. However, a major concern is the accuracy of these low-cost PM sensors. This chapter will introduce working principles of low-cost PM sensors, together with calibration methods and calibration metrics used in previous research to evaluate the performance of low-cost PM sensors. We will also introduce how these low-cost PM sensors have assisted pollution mapping in real-life applications. Finally, the challenges regarding the calibration and application of low-cost PM sensors have been discussed.

Keywords: low-cost pm sensors, sensor application, pollution mapping, calibration methods, calibration metrics

4.1 Introduction

New designs for low-cost particulate matter (PM) sensors, new commercial products, and an accompanying number of new publications show the trending interest in this area. The deployment of low-cost PM sensor networks, together with their pros and cons, has also been discussed in recent works [1–3]. Compared to conventional particulate matter (PM) monitoring techniques, the price advantage and minimal maintenance of low-cost PM sensors make them a promising supplement to current monitoring methods. They can enhance the spatiotemporal resolution of

Jiayu Li, Pratim Biswas, Department of Bioproducts and Biosystems Engineering, University of Minnesota, Twin Cities, 1390 Eckles Ave, BioAgEng Building, St. Paul, MN 55108, USA

https://doi.org/10.1515/9783110729481-004

pollution mapping, improving the accuracy of personal exposure estimation and validation of the PM transport models. Estimating of the personal PM exposure accurately can benefit epidemiologic studies by identifying the adverse health effects of PM. Understanding and improving PM transport models can effectively control and even prevent pollution events. These promising applications explain the recent extensive studies of low-cost PM sensors. As shown in Figure 4.1, publications related to air quality sensors increased steadily since 2013.

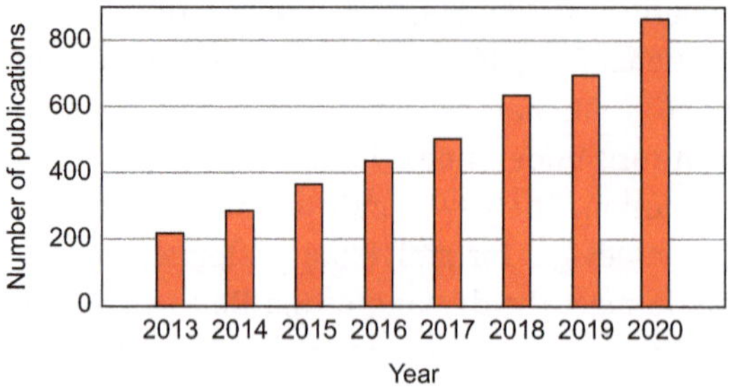

Figure 4.1: (a) The number of papers published each year related to "air quality sensors," data from Web of Science.

Studies related to low-cost sensors basically focus on their calibration and application. The calibration studies evaluate sensors' performance by comparing them with reference instruments, while the application studies focus on pollution mapping and personal exposure estimation. Several reviewers have summarized studies related to low-cost PM sensors [4–7]. Kumar et al. (2015) generally explained the motivations for the rising topic, and reviewed concerns of the reliability, sensitivity, selectivity, and durability of low-cost sensors [4]. Rai et al. (2017) concretely summarized literature on the performance of several different types of low-cost PM sensors, and also analyzed possible environmental factors and aerosol properties that could bias their performance [5]. Morawska et al. (2018) analyzed 17 large ongoing funded research studies on low-cost PM sensors, and summarized the major concerns regarding sensor calibration and application [6]. Synder et al. (2013) highlighted that low-cost sensors can improve existing air pollution monitoring capabilities and inspire innovative applications [7]. Several other reviews illustrate the current limitations and the future of low-cost air quality sensors [8–10].

In this chapter, we further review the studies of low-cost PM sensors, and focus in detail on their working principles, calibration methods, calibration metrics, and application scenarios. The working principles of several low-cost PM sensors are demonstrated, using schematics from previous studies. Also sensors for measuring the PM with other techniques are briefly discussed. Calibration methods and metrics are summarized and compared. These methods and tools include regression or correlation, the non-parametric Wilcoxon signed-rank test, the ranking method, the Bayesian information criterion, average slope and individual slope methods, and, finally, the machine learning method. Calibration metrics, parameters for evaluating the performance of low-cost PM sensors, include the limit of detection (LOD), the correlation coefficient of linear regression, bias and precision, the coefficient of variation (CV), and the detection efficiency. Finally, innovative applications of low-cost PM sensors in field measurements or personal exposure estimation are discussed. Also, for corresponding cases, we introduce several spatial analytic methods, illustrating the use of the coefficient of divergence (COD), land use regression (LUR), and several spatial interpolation methods. One thing worth noting is that the term "low-cost PM sensor" generally refers to both electrical sensing modules (e.g., popular models from Sharp, Shinyei, Samyoung, Oneair, and Plantower), together with low-cost PM monitors based on sensing modules. To make the sensing module functional, circuit board design, programming, and calibration are necessary to establish the relationship between electrical signals (current, voltage, or pulse width) and PM concentrations. For low-cost PM monitors, electrical sensing modules have been integrated with data acquisition and storage systems before being distributed to users, and they have been calibrated and tested. Compared to the PM sensing module alone, the assembled monitors' prices are higher, but these monitors are advertised with enhanced data quality and stability due to improved algorithms and advanced factory calibration. Occasionally, these monitors have even been chosen as reference instruments to calibrate low-cost sensors. For convenience, here we still use the general term "low-cost PM sensors" for both sensing modules and low-cost PM monitors.

4.2 The working principles of low-cost PM sensors

Low-cost PM sensors, operating on basic optical principles, determine the PM concentration level by measuring the intensity of light scattered by particles. Basically, there are two types of these sensors: nephelometer-type sensors and optical particle counter (OPC)-type sensors. In a nephelometer-type sensor, particles pass through the sensing volume almost simultaneously in a cloud, and the particle concentration is determined by the total scattered light intensity registered by a photodetector. In an OPC-type sensor, when a single particle passes the sensing volume, the scattered light generates a pulse on the photodetector. The number and the intensity of pulses are proportional to PM's number concentration and size, respectively. The working principles of several popular types of low-cost PM sensors are shown in Figure 4.2. Only the Plantower sensors in Figures 4.2(g) and 4.2(h) are OPC-type sensors; the rest are the nephelometer-type sensors. Apart from the commercial designs shown in Figure 4.2, several studies have proposed new designs for the low-cost PM sensors, focusing on eliminating the effect of the particles' refractive index or enhancing the sensors' accuracy [11–14].

The light scattering techniques are often considered capable of measuring only particles larger than 0.3 µm. This conclusion is accurate for OPC-type sensors, since the pulse signal generated by a very small particle will be buried in the noise. For the nephelometer-type sensors, individual small particles cannot generate intensive signals. However, if the number of particles and, consequently, the particle concentration are high enough, particles can still generate a detectable response, since the totally scattered light intensity is also related to the number of concentration.

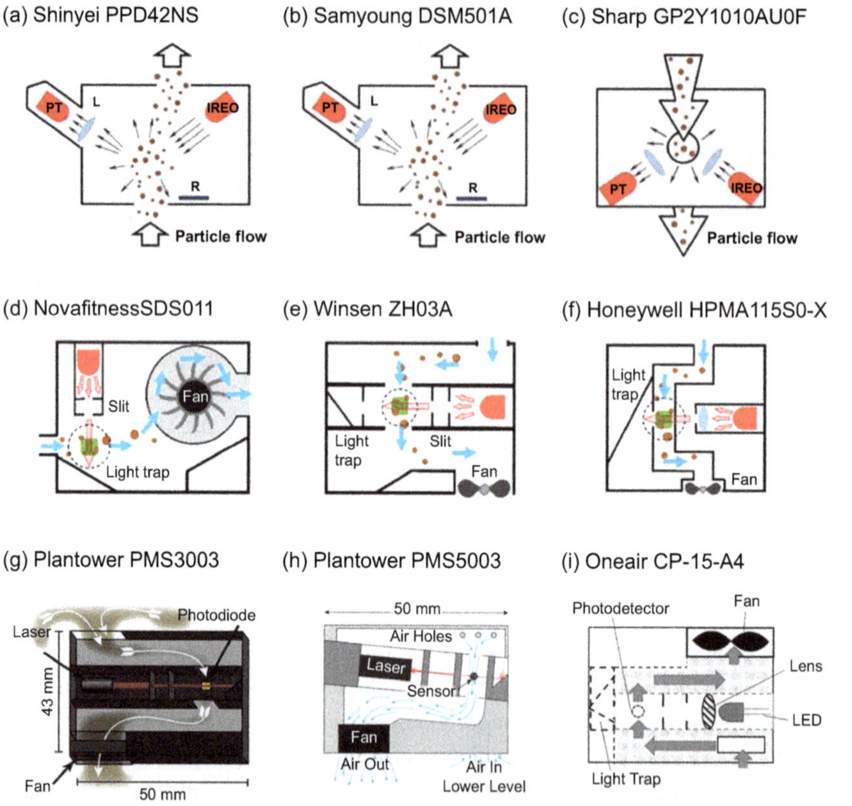

Figure 4.2: The working principles of (a) Shinyei PPD42NS, (b) Samyoung DSM501A, (c) Sharp GP2Y1010AU0F, (d) NovafitnessSDS011, (e) Winsen ZH03A, (f) Honeywell HPMA115S0-X, (g) Plantower PMS3003, (h) Plantower PMS5003, and (i) Oneair CP-15-A4. The figures are from the following studies – (a–c): Wang et al. .[30], (d–f): Hapidin et al. [32], (g): Kelly et al. [34], (h): Sayahi et al. [59], and (i): Liu et al. [31].

Apart from optical sensors, other types of low-cost PM sensors are receiving attention. Intra et al. (2013) presented a design based on unipolar corona charging and electrostatic detection of charged particles [15]. Volckens et al. (2017) designed a time-integrated filter sampler with an ultrasonic piezoelectric pump to drive flow, together with a cyclone to select particles of a certain size range [16]. Surface acoustic wave sensors can detect PM concentration by measuring the resonant frequency change after particles deposit on the sensing area, which interferes with the propagation of acoustic waves [2, 17–19]. Budde et al. (2013) designed an add-on PM detector component for smart phones, using the flashlight and camera as the light source and photo detector [20]. Snik et al. (2014) proposed an attachable component that assists smart phones for aerosol optical thickness measurement [21]. Du et al. (2018) developed the a PM sensor based on a CMOS (complementary metal oxide semiconductor) imager and an electrostatic particle collector [22]. A similar design was also reported by Carminati et al. [23]. Yang et al. (2018) synthesized a layer of polypyrrole sensing nanofilm on a photonic crystal fiber [24]. Particles deposited on the sensing nanofilm change its refractive index, indicating the PM concentration level change [24]. The feasibility of using photonic and microelectromechanical resonators for detecting particles or viruses has also been discussed [25]. Recently, a piezoelectric microelectromechanical resonator, together with a low-cost circuit, was proposed as a new low-cost PM sensor [26]. The quartz crystal microbalance (QCM) for measuring PM mass concentration, based on the frequency shift caused by particle deposition. The miniaturized devices based on QCM have been developed recently [27, 28].

Although these new innovative designs have an intriguing future, the low-cost PM sensors operating on optical principles are still the dominant type, for several reasons. First, theories of the interactions between light and particles are maturely developed. At the same time, many research-grade PM measurement instruments are also based on optical principles, and researchers are familiar with these instruments. Therefore, it is easy for them to shift from using conventional instruments to low-cost PM sensors operating on a similar principle. To prove the reliability and stability of the innovative designs, further effort is still needed. Second, there are also many concerns about their cost and fabrication procedures. For some innovative design, although the sensing unit is low-cost, the signal processing and detection components are expensive. In summary, the nephelometer-type and OPC-type designs are still the most widely used because they are compact and are easily integrated with other systems. These devices operate on familiar principles and are conveniently deployable.

4.3 Calibration methods overview

Laboratory calibrations and field calibrations are the foundations of low-cost PM sensors' applications [29–35]. In a laboratory calibration, environmental factors and aerosol properties can be controlled. Environmental factors include the temperature, relative humidity, and ventilation rate. For aerosols, ultrafine particles can be generated from atomizers, and micron-sized particles can be generated from dust dispensers. A few studies have also used common residential or industrial PM sources (e.g., cookstoves and cigarettes) to mimic practical situations in laboratories. Salt particles, sucrose particles, cigarette emissions, welding fumes, and Arizona road dust have been used in laboratory calibrations. The size distribution and composition of aerosols can be controlled well in laboratory experiments, which benefits the analysis of low-cost PM sensors' dependence on these variables. In laboratory calibration, low-cost sensors are usually compared with the scanning mobility particle sizer (SMPS), aerodynamic particle sizer (APS), GRIMM, and TSI portable-series instruments (e.g., SidePak, DustTrak, and PTrak).

Field calibration focuses more on the performance of low-cost PM sensors under uncontrolled and dynamic environments, and can be conducted in residential or outdoor environments. In a residential environment, the sensors' responses to routine PM emission events (e.g., cookstove emissions, woodworking shop operations, and incense burning) can be studied. Outdoor calibration focuses on agreement between low-cost PM sensors and federal reference methods, including the gravimetric method, the β-attenuation analyzer, and the tapered element oscillating microbalance (TEOM). Outdoor emissions, especially urban traffic emissions, have been characterized in several studies. In field calibrations, the PM composition and concentration levels can be highly dynamic. Thus, the time domain is usually longer than in laboratory calibrations to collect enough data over a whole concentration range. Although the performance of low-cost PM sensors in field calibration may not be as good as that in laboratory calibration, the results from field calibrations are closer to the real situation, and field calibration is a good method to examine the reliability, durability, and longevity of low-cost PM sensors.

Here we first discuss several calibration methods that have been used in previous studies, including linear regression or correlation, the reduced major axis method, Bayesian information criterion, non-parametric Wilcoxon signed-rank test, the average slope and individual slope method, and machine learning method. Then, methods to correct sensors' performance for the effects of temperature and relative humidity are briefly discussed.

Although there is a slight difference between correlation and linear regression, they are the most common methods for calibrating low-cost PM sensors. The sensors' outputs are plotted against the outputs from reference instruments, and a fitted equation is used to optimize the accuracy of the sensors' outputs. The correlation coefficient, R, is a statistic measuring the degree or strength of linear correlation. In different studies on evaluating sensors' performance, R is referred as the r coefficient, Pearson's product-moment r, or Pearson's correlation coefficient. R values, typically given with two decimal places, range from -1 for a strong negative correlation, through 0 for a no or a weak correlation, to $+1$ for a strong positive correlation. The value of R^2 is rescaled to 0 to 1, describing purely the strength of the correlation. Several authors have explained that combining the hypothesis test (p-value significance test) with the r or

R^2 value is a more rigorous method for judging the relationship between two variables [36–38].

If not otherwise specified, the linear regression or correlation is usually based on the least square method. The reduced major axis method, in addition to the least square method was used in several studies, to calculate the correlation coefficient, slope, and intercept [30, 32]. The assumption of the least square method is that the independent variables are measured accurately [39, 40]. Therefore, if we are calibrating low-cost sensors against a reference instrument and are very confident about the results from the reference instrument, the least squares method is appropriate. However, in situations where the accuracy of the reference instrument is underdetermined, or when comparing the low-cost sensor against another low-cost sensor, the reduced major axis method is more applicable, because within this method, the error of both the dependent and independent variables is considered [40, 41].

To improve sensors' performance by including more variables in the model (for example, the relative humidity and temperature), the linear correlation or regression may not be adequate. Gao et al. (2015) used both the Bayesian information criterion (BIC) and the standard error of regression to evaluate fitted models that included temperature and humidity as variables [42]. The BIC method can prevent overfitting by introducing a penalty term that reflects the number of free parameters in the model [43–46]. The standard error of regression, also known as the standard error of the estimation, evaluates the difference between observed and model-predicted values. For complicated models, increasing the number of free parameters, for example, by including more variables or fitting with higher orders, will reduce the standard error of regression. However, it will also lower the BIC number due to the penalty term [47]. By combining these two methods, an optimal predictive model can be selected with the minimal discrepancy from the observations, without overfitting.

The non-parametric Wilcoxon signed-rank test was used by Zikova et al. (2017) in evaluating the performance of Speck sensors [48]. This method, also known as Mann-Whitney *U* test or Mann-Whitney-Wilcoxon test, can be used to examine whether the sensor data and the reference instrument data are from the same population. If they are from a same population, then the quality of the sensor is satisfactory and it can be a replacement for the reference instrument. Unlike most statistical methods (e.g., Student's *t*-test), which require the assumption of a normal distribution, the Wilcoxon signed-rank test is simple and intuitive. It does not require assumptions about the distribution of the data [49, 50]. However, it only qualitatively demonstrates whether a hypothesis can hold, and is inadequate to quantify the magnitude of any effect. Another rank order analysis method was used by L. R. Crilley for evaluating the variability of 14 Alphasense OPC-N2s over a period of time [51]. The PM measurements were ordered from the highest to the lowest, after being normalized to the median concentration at the start of the analysis. Compared to the pair-wise correlation, this method can show the dynamics of the variation, such as offset, as a function of time. The offset drift, or the temporal consistency of each sensor, can also be demonstrated by this method. Ideally, sensors which report the higher PM concentrations than peer sensors initially are supposed to report higher concentration at the end of the measurement period as well, representing no drift or the same degree of drift.

The average slope and individual slope method was used to guide the deployment of a low-cost sensor network in a heavy-manufacturing site, for convenience in calibrating the multiple sensors [52]. A major concern in the deployment of low-cost sensors is their unpredictable data quality. Repeated calibration has been recommended to enhance the data quality; however, it is time consuming and inconvenient for tens of sensors in a field deployment. In studies mentioned above, the average slope method was used for selecting sensors with similar slopes in the calibration stage. Then, in the field deployment, the reference instrument was collocated with only several of the selected sensors, and a universal field calibration factor was applied to all the selected sensors.

Machine learning, as a popular concept in computer science, has also been used for sensor calibration. A feedforward Neural Network has been used in the calibration of the Plantower PMS7003 [53]. An artificial neural network has been used to predict the PM distribution in a woodworking shop [54]. Zimmerman et al. (2018) compared three calibration methods, including the laboratory univariate linear regression, the empirical multiple linear regression, and the machine-learning method (random forest) to calibrate different gas sensors, and these methods should be considered in the calibration of low-cost PM sensors [55].

The influence of temperature and relative humidity on the performance of low-cost PM sensors has been studied in the field and lab studies. Some studies have indicated that the influence of temperature was a negligible effect on sensors' performance [30, 56]. Several other studies have concluded that high RH may bias the performance of low-cost PM sensors in both laboratory calibration and field evaluation [57, 58]. However, still other studies, especially in field evaluations, indicate that the influence of relative humidity is negligible [59]. There have been attempts to eliminate the influence of relative humidity and temperature by including empirical equations, fitted equations, or hygroscopic growth factors in a more complicated model to calibrate the low-cost PM sensors [42, 60–65]. Compared to research grade instruments, low-cost PM sensors lack temperature and humidity control components, and thus changes of shape, size, phase (solid to liquid or liquid to solid), and optical properties of particles under high relative humidity may bias their performance [66, 67]. The influence of relative humidity has been extensively studied by atmospheric scientists [68–70]. The influence of relative humidity also depends on particle surface properties and compositions, which may explain why, in several field studies, the relative humidity did not show a significant influence on sensors' performance [66]. Further study is needed to explore in detail how environmental factors can influence the sensors' performance and how to correct such bias.

As mentioned above, researchers have tried several different calibration methods to improve the performance of low-cost PM sensors. It has been very controversial whether all studies should follow the same calibration methods for calibrating different kinds of sensors, so that the results from different reports can be comparable. However, a concern here is that such a standard guideline might discourage exploring and applying the new statistical methods for the sensor calibration. In addition, other issues arise. First, the current OPC-type sensors and nephelometer-type sensors follow the same procedures for the calibration. Considering the differences in their working principles and measurement metrics, the calibration methods may need reevaluation and modification. Second, the criterion for calibrating low-cost sensing modules (solely electrical component) and low-cost PM monitors (calibrated and tested before being distributed to users) is worth further discussion. For low-cost sensing modules, the focus of calibration is whether a good linearity can be established; however, for low-cost PM monitors, the agreement with reference instruments might be more important. For the sensing modules, calibration is necessary to establish the relationship between the electrical signals and $PM_{2.5}$. For low-cost PM monitors, since they already report $PM_{2.5}$, the bias and deviation should be the focuses, instead of the correlation. Therefore, it is necessary to distinguish between sensing modules and low-cost PM monitors, since their calibration metrics and methods are inherently different.

4.4 Calibration metrics

Calibration metrics are parameters whose values are calculated from the calibration procedures used to evaluate the performance of low-cost PM sensors. For example, the correlation coefficient of the linear regression is a common parameter to evaluate the linearity of low-cost PM sensors. Similar metrics include the limit of detection (LOD), the bias and precision, and the coefficient of variation (COV).

The R^2 value determined from the linear regression is a primary parameter to evaluate the linearity of low-cost PM sensors. Details related to the linear regression have been mentioned in the last section. The R^2 values of low-cost PM sensors previously studied, as summarized by Rai et al. [5], are presented in Figure 4.3. In the literature, R^2 values are reported for different PM sources under various test environments. The maximum and the minimum of R^2 values for several types of low-cost PM personal monitors are summarized in Figure 4.4. The major components of several low-cost PM monitors (see Figure 4.4) are sensing modules mentioned in Figure 4.3. For example, the major component of the AirAsure is the Sharp GP2Y1010AU0F, and the major component of the PurpleAir is the Plantower PMS series low-cost sensing module. Since the tests in Figures 4.3 and 4.4 were not conducted following the same methodology and guidelines, the reported R^2 values could vary with different test conditions, and results may not be directly comparable. However, the trend is basically the same: the R^2 value from the laboratory calibration ($R^2 > 0.6$) is better than that from field calibration ($R^2 > 0.4$).

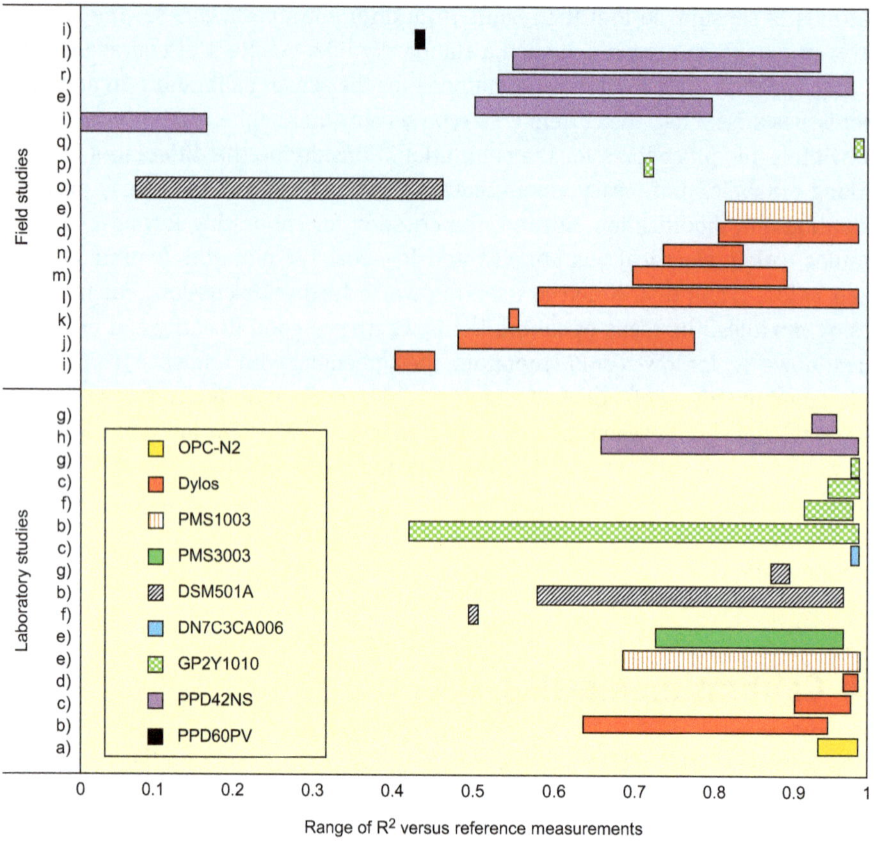

Figure 4.3: The R^2 values of low-cost PM sensors, summarized by Rai et al. [5].

The limit of detection (LOD) is the lowest detectable concentration that significantly stands out from the background noise. Low-cost PM sensors are considered to report reliable and meaningful data, only when the concentration exceeds the LOD. Equation (4.1) for calculating the LOD is given below, where k and σ_{blk} represent the slope from the fitted model and the standard deviation of low-cost PM sensors under a particle-free background, respectively [30, 71]. Knowing the LOD before the deployment is necessary to produce reliable data, especially for atmospheric measurement.

$$LOD = 3\sigma_{blk}/k \tag{4.1}$$

Equation (4.2a) shows the bias defined by the National Institute for Occupational Safety and Health (NIOSH), where C_{LS} and C_{rf} represent the concentrations measured by low-cost sensors and reference instruments [33, 72, 73]. NIOSH bias, also

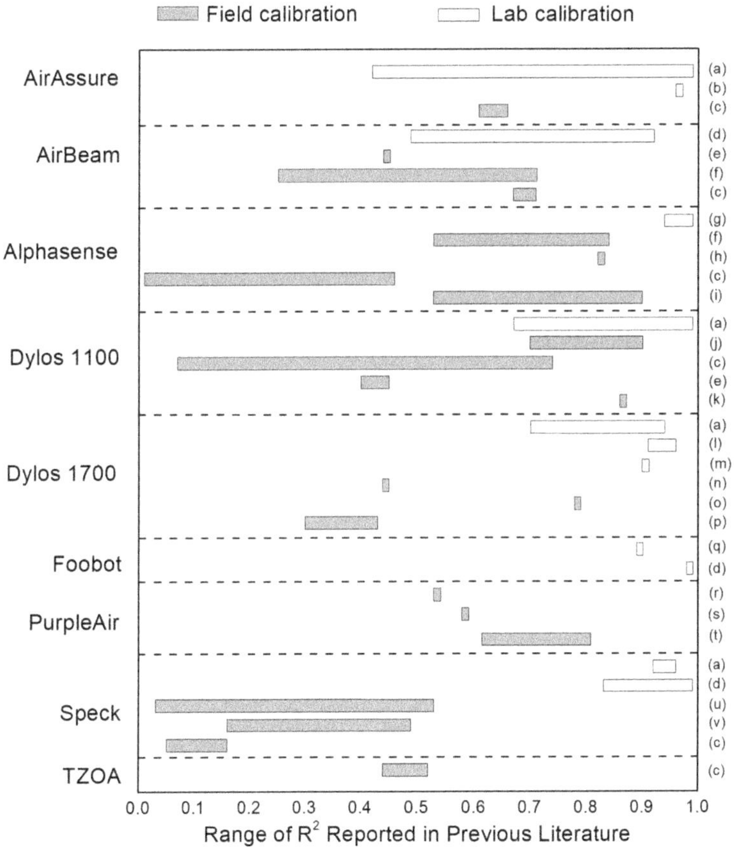

Figure 4.4: Reported R^2 values of low-cost PM personal monitors. The data are from the following studies – (a) Manikonda et al. [75], (b) Wang et al. [30], (c) Feinberg et al. [120], (d) Sousan et al. [121], (e) Jiao et al. [122], (f) Mukherjee et al. [123], (g) Sousan et al [33]., (h) Gillooly et al. [124], (i) Crilley et al. [51], (j) Steinle et al. [95], (k) Semple et al. [125], (l) Sousan et al. [33], (m) Semple et al. [126], (n) Jovašević-Stojanović et al. [127], (o) Han et al. [128], (p) Franken et al. [129], (q) Moreno-Rangel et al. [130], (r) Sayahi et al. [59], (s) Malings et al. [64], (t) Malings et al. [64], (u) Zikova et al. [48], and (v) Zikova et al. [48].

known as the percent difference, evaluates the percent of error of low-cost sensors' output compared to reference instruments. Zikova et al. (2017) and Sousan et al. (2018) have used this method to evaluate the performance of Speck monitors and Sharp sensors respectively [48, 52]. Sayahi et al. (2019) used a similar definition, referred to as the normalized residual, to evaluate Plantower sensors [59]. The EPA specifies the bias of low-cost PM sensors by eq. (4.2b), which gives the average of the percent difference in k different measurements [72]. Both NIOSH bias and EPA bias have been recommended to be within ± 10%.

$$NIOSH\ bias\ (d_i) = \frac{C_{LS} - C_{rf}}{C_{rf}} \times 100\% \tag{4.2a}$$

$$EPA\ bias\ (b_i) = \frac{1}{k}\sum_{i=1}^{k} \frac{C_{LS} - C_{rf}}{C_{rf}} \times 100\% = \frac{1}{k}\sum_{i=1}^{k} d_i \times 100\% \tag{4.2b}$$

The measurement precision parameter reflects the stability and repeatability of low-cost PM sensors at a certain concentration level. There are several different definitions of measurement precision. For evaluating the repeatability and stability of low-cost PM sensors for a fixed concentration level Long et al. (2014) defined measurement precision, as shown in eq. (4.3a), where P_1, P_2, and P_3 represent three individual measurements of low-cost PM sensors for the same concentration level [74]. Manikonda et al. (2016), Zikova et al. (2017), and Zamora et al. (2018) used a similar definition of precision that involves the difference and mean of the sensor' and reference instruments' outputs, as shown in eqs. (4.3b and 4.3c) [48, 61, 75]. Manikonda et al. (2016) used the normalized root mean square error to quantify the difference of paired data, as shown in eq. (4.3b) [75]. In this equation, n is the number of data pairs in a period of the measurement, and P_i and C_i represent the paired data from two low-cost PM sensors, respectively. Zikova et al. (2017) and Zamora et al. (2018) used the same definition (eq. (4.3c)) and referred to the method as "unbiased variance estimate" and "relative precision error," respectively [48, 61]. Compared to eq. (4.3a), eqs. (4.3b and 4.3c) might be more practical for calibration since they do not require a fixed concentration level, and they consider the precision values with respect to concentration levels by normalizing them to average measurement results.

$$Precision = \sqrt{\frac{1}{2}\left[\sum_{i=1}^{3} P_i^2 - \frac{1}{3}\left(\sum_{i=1}^{3} P_i\right)^2\right]} \tag{4.3a}$$

$$Precision\ of\ measurement = \frac{\sqrt{\frac{1}{n}\sum_{i=1}^{n}(P_i - C_i)^2}}{\frac{1}{n}\sum_{i=1}^{n}(P_i + C_i)/2} \tag{4.3b}$$

$$Precision = \frac{|P_i - C_i|}{P_i + C_i} \times 100\% \tag{4.3c}$$

The coefficient of variation (CV), another parameter for evaluating the precision of low-cost PM sensors, is defined by eq. (4.4), where σ and μ represent the standard deviation and the mean of measurements, respectively. CV measures the degree of variation, indicating the dispersion of data points around the mean value. Sousan et al. (2016) and Zamora et al. (2018) have used this parameter to evaluate the performance of the Alphasense OPC-N2 and Plantower PMS A003 respectively [33, 61]. Several other studies have used CV to evaluate different types of sensors [76]. The satisfactory CV value of less than 10% is taken as

$$CV = \frac{\sigma}{\mu} \tag{4.4}$$

Apart from the parameters mentioned above, other statistical measures can be used to evaluate the performance of low-cost sensors. Examples include the median, mean, mode, the 25th and 75th percentiles [51], and the mean relative standard deviation [57]. All these statistical measures quantify accuracy and repeatability of low-cost sensors from different perspectives. However, current studies and guidelines have limitations. First, the criterion for "good" performance is vague. In the methods mentioned above, only reference values for CV and bias are given by NIOSH and EPA. More criteria are needed, for example, in what range the reference value of R^2 can be called a "good" PM sensor. Several guidebooks discuss standard procedures and guidelines to calibrate low-cost sensors, led by the EPA and air quality sensor performance evaluation center (AQ-SPEC), and we expect more discussion on this subject [77–79]. Second, the performance of low-cost PM sensors varies with the particle size distribution, composition, and testing environment, which makes the results from different reports difficult to compare. A guideline for specifying the test conditions would be helpful in this field. Third, several different parameters are reported by low-cost PM sensors, including the number concentration, mass concentration, and size distribution. Sensors reporting number concentration and mass concentration have been evaluated by the statistical methods mentioned in this section, however, there are limited options for quantifying accuracy of the size distribution data. Normally, the size distributions from low-cost PM monitors are plotted together with those from reference monitors. Also, sometimes, detection efficiency was calculated to quantify the performance. More discussion is needed to evaluate size distribution measurements from different perspectives.

4.5 Applications

The superiority of low-cost PM sensors, their price advantage, portable size, and moderate accuracy have made them a good supplement to current monitoring stations. Several studies have shown that the spatial variation cannot be neglected, even over a kilometer scale [80, 81], and such small-scale heterogeneity is important for accurately quantifying the personal exposure level [82]. Here we present several examples of field deployment of low-cost PM sensors, together with related spatial analysis methods. An important topic in the sensor deployment is an application of the interpolation method for predicting the PM concentration at locations with no measurements, known as pollution mapping. Common interpolation methods were also discussed in this section.

Low-cost PM sensors have been innovatively applied in industry or daily life. Low-cost sensors have been used to examine the relationship between different sources and PM concentration levels [83]. Dylos sensors were used to evaluate the pesticide off-target drift of agricultural tower sprayers [84]. A Novafitness low-sensor was used to evaluate the emissions of a surface filter [85]. Other than estimating the personal exposure [86] and mapping pollution distribution [87], applications also include characterizing households emissions [88, 89], cigarette emissions [90], and industrial factory emissions [54, 91]. Low-cost PM sensors can also contribute to the construction of smart cities that provide personal exposure estimation with better accuracy [92–94].

Apart from deploying fixed sensors to enhance the spatial resolution, several studies have involved mobile sensing nodes. When combined with a data logging system (e.g., on a microSD card) or position logging system (e.g., GPS), low-cost sensors can be used to refine the assessment of personal exposure [20, 95–99]. At the indoor scale, an ultrasonic indoor positioning system has been used to position mobile low-cost PM sensors for the indoor exposure estimation [100]. These sensors have also been integrated with unmanned aerial vehicles (e.g., drones) for outdoor vertical measurement [101, 102]. Furthermore, a low-cost robot carrying the low-cost PM sensor has been tested for remote sampling or autonomous sampling [103].

The coefficient of divergence (COD), defined by eq. (4.5), quantifies the level of heterogeneity between two places, where $x_{i,\,j}$ and $x_{i,\,k}$ are the ith measurement at location j and k respectively [104]. Normally, the COD value smaller than 0.2 represents no significant difference between the measurements at two different locations, indicating homogeneity. COD values larger than 0.2 represents increasing heterogeneity [104, 105]. Zikova et al. (2017) and Saha et al. (2019) had used this method to examine, respectively, the PM distribution with 25 Speck monitors in New York and 32 RAMP (real-time, affordable, multi-pollutant) monitors in Pittsburg respectively [48, 81]. Reece et al. (2018) used this method in a field campaign in Puerto Rico to analyze the spatiotemporal distribution of $PM_{2.5}$ and NO_2 [76]. Using the COD of different species, Saha et al. (2019) found that ultrafine particles and $PM_{2.5}$ respectively demonstrated higher and lower heterogeneity [81].

$$COD_{j,k} = \sqrt{\frac{1}{p}\sum_{i=1}^{p}\left[\left(x_{ij} - x_{ik}\right)/\left(x_{ij} + x_{ik}\right)\right]^2} \tag{4.5}$$

Different types of interpolation methods have been used for indoor and outdoor pollution mapping employing low-cost sensors. Zikova et al. (2017) used the inverse squared-distance weighing interpolation (IDW) to predict the outdoor PM distribution with 25 Speck monitors [48]. Li et al. (2018) mapped the spatiotemporal PM distribution in a woodworking shop with 8 Sharp GP2Y sensors by Kriging interpolation and artificial neural network method [54]. In two consecutive studies, For Melbourne (Australia), Rajasegarar et al. (2014) used a Bayesian maximum entropy (BEM) method to map the PM distribution obtained using GP2Y sensors [106, 107]. In environmental studies, around 30–40 types of spatial interpolation methods have been deployed Several reviews presented well-grounded summaries explaining how these methods differ and how to quantify the accuracy of the interpolation results [108–111]. The major difference among various types of interpolation methods is the weight assigned to measured data for predicting the concentration at the unsampled locations. Methods used to assess the accuracy of the measurement can also be used to optimize the sampling locations.

Land use regression (LUR) is another important method applied for interpreting the results from low-cost PM sensors [112]. LUR also uses data at sampled locations to predict the PM values at the unsampled locations. However, a major difference that distinguishes LUR from other interpolation methods is the involvement of additional predictor variables, for example, land use, traffic, population density, physical geography, and meteorology. Including these additional predictor variables shows that the PM concentration is not only a function of the location. Another large difference between the spatial interpolation methods discussed above and the LUR is the restriction of measured datasets. For Kriging interpolation, the highest PM concentration belongs to the measured datasets. However, for the LUR, the highest PM concentration may not be the highest measured value, because the results are influenced by multiple variables.

Several questions demand further attention. First, the differences between the indoor and outdoor pollution mapping need to be highlighted. Indoor pollution events are highly dynamic. They can be changed within several seconds due to complicated ventilation conditions. A special question is how to adjust the sampling locations and intervals to meet the requirements of indoor and outdoor pollution mapping. A second question is related to the strategy of measurement. Most monitoring stations are located heterogeneously, concentrated in and around metropolitan areas or in industrial areas. The question of how this sparseness may undermine the accuracy of pollution mapping should be answered in the future studies. Optimizing the locations of low-cost PM sensors is another potential topic related to the efficiency and effectiveness of measurements. The third question is how best to connect the pollution mapping results with studies in other fields. The pollution mapping results from conventional methods have already been used in epidemiologic studies. Low-cost PM sensors can indeed provide data with better spatiotemporal resolution; however, the reliability and accuracy of the data remain as concerns.

4.6 Challenges

We have summarized the methods and metrics used in previous studies to evaluate the performance of low-cost PM sensors. These methods can demonstrate the advantages and limitations of each type of sensor. Characterizing these sensors thoroughly will benefit their deployment in field studies. Low-cost PM sensors have demonstrated acceptable accuracy and stability in the calibration and characterization, which demonstrates great potentials in various applications for mapping the pollution distribution and quantifying personal exposure. However, there are several challenges which still generate concerns in the current studies.

There was distinguished a set of parameters that may bias sensors' performance, leading to overestimating or underestimating in mass concentrations. In the previous section, we have also mentioned that environmental parameters (e.g., relative humidity and temperature) and PM properties (e.g., size distribution and optical properties) may all challenge the accuracy of the low-cost PM sensors. Either establishing the models with more parameters or improving sensor structures can achieve a better accuracy. Several studies mentioned above have built different models to correct the bias caused by the relative humidity and temperature. However, a big concern is whether the models are universal, applicable to all or most of the scenarios. To establish a universal model, more fundamental studies are needed. At the same time, there are limited studies of improving the sensors' structure for better performance. Therefore, how to improve sensors' performance for an accurate estimation of the PM mass concentration needs further efforts.

Strengthening of sensors' performance by lowering the LOD and cut-off size. The LODs were not always reported in the literature as correlation coefficients. However, they are vital for judging whether a specific type of sensors is appropriate for the deployment. A further lowering of the LOD with either advancing algorithms or sensor structure improvement will benefit the sensors' deployment in the field measurement. The cutoff size for OPC-type sensors is approximately 300 nm, which does not include ultrafine particles yet. Furthermore, ultra-fine particles have showed a stronger mobility and have demonstrated a more heterogeneous distribution than larger particles. Therefore, the lowering of the cut-off size for both OPC-type and nephelometer-type PM sensors is necessary and practical.

Lowering the cost for maintaining a sensor network. Although the cost and maintenance of a single PM sensor is low, maintaining the sensor network will be a different story. Although the maintenance requirement for low-cost PM sensors is lower compared to conventional methods, it still cannot be ignored. Identifying the malfunctioning sensors and repairing them will be difficult for the sensor network with more than a hundred units. How to enhance the stability and robustness of low-cost PM sensors to realize zero-effort maintenance will benefit field applications.

Exploring of different applications of low-cost PM sensors. The data from low-cost sensors have been used for regional pollution mapping and exposure estimation [113–116]. In a few studies, robots and drones have been used with low-cost PM sensors to realize the autonomous measurement. Some attempts have been made to combine the low-cost PM sensors with remote sensing or ground measurement [117–119]. More studies are expected to explore the possibilities of using the sensor data in different applications and different scenarios.

References

[1] Amegah, A.K. (2018). Proliferation of low-cost sensors. What prospects for air pollution epidemiologic research in Sub-Saharan Africa? *Environmental Pollution* 241:1132–1137.

[2] Thomas, S., Cole, M., Villa-López, F.H., Gardner, J.W. (2016). High frequency surface acoustic wave resonator-based sensor for particulate matter detection. *Sensors and Actuators. A, Physical* 244:138–145.

[3] Castell, N., Dauge, F.R., Schneider, P., Vogt, M., Lerner, U., Fishbain, B., Broday, D., Bartonova, A. (2017). Can commercial low-cost sensor platforms contribute to air quality monitoring and exposure estimates?. *Environment International* 99:293–302.

[4] Kumar, P., Morawska, L., Martani, C., Biskos, G., Neophytou, M., Di Sabatino, S., Bell, M., Norford, L., Britter, R. (2015). The rise of low-cost sensing for managing air pollution in cities *Environment International* 75:199–205.

[5] Rai, A.C., Kumar, P., Pilla, F., Skouloudis, A.N., Di Sabatino, S., Ratti, C., Yasar, A., Rickerby, D. (2017). End-user perspective of low-cost sensors for outdoor air pollution monitoring. *Science of the Total Environment* 607:691–705.

[6] Morawska, L., Thai, P.K., Liu, X., Asumadu-Sakyi, A., Ayoko, G., Bartonova, A., Bedini, A., Chai, F., Christensen, B., Dunbabin, M. (2018). Applications of low-cost sensing technologies for air quality monitoring and exposure assessment: How far have they gone?. *Environment International* 116:286–299.

[7] Snyder, E.G., Watkins, T.H., Solomon, P.A., Thoma, E.D., Williams, R.W., Hagler, G.S., Shelow, D., Hindin, D.A., Kilaru, V.J., Preuss, P.W. (2013). The changing paradigm of air pollution monitoring. In ACS Publications.

[8] Koehler, K.A., Peters, T.M. (2015). New methods for personal exposure monitoring for airborne particles. *Current Environmental Health Reports* 2(4):399–411.

[9] Thompson, J.E. (2016). Crowd-sourced air quality studies: A review of the literature & portable sensors. *Trends in Environmental Analytical Chemistry* 11:23–34.

[10] Kumar, P., Skouloudis, A.N., Bell, M., Viana, M., Carotta, M.C., Biskos, G., Morawska, L. (2016). Real-time sensors for indoor air monitoring and challenges ahead in deploying them to urban buildings. *Science of the Total Environment* 560:150–159.

[11] Njalsson, T., Novosselov, I. (2018). Design and optimization of a compact low-cost optical particle sizer. *Journal of Aerosol Science* 119:1–12.

[12] Dong, M., Iervolino, E., Santagata, F., Zhang, G., Zhang, G. (2016). Silicon microfabrication based particulate matter sensor. *Sensors and Actuators. A, Physical* 247:115–124.

[13] Qiao, Y., Tao, J., Zhang, Y., Qiu, J., Hong, X., Wu, J., Chen, C.-H. (2019). Sub-micro particle matter detection for metal three dimensional printing workshop. *IEEE Sensors Journal* 2151–2154.

[14] Li, X., Iervolino, E., Santagata, F., Wei, J., Yuan, C., Sarro, P., Zhang, G. (2014). In *Miniaturized particulate matter sensor for portable air quality monitoring devices*, SENSORS, 2014 IEEE; IEEE: 2014; pp 2151–2154.

[15] Intra, P., Yawootti, A., Tippayawong, N. (2013). An electrostatic sensor for the continuous monitoring of particulate air pollution. *Korean Journal of Chemical Engineering* 30(12): 2205–2212.

[16] Volckens, J., Quinn, C., Leith, D., Mehaffy, J., Henry, C.S., Miller-Lionberg, D. (2017). Development and evaluation of an ultrasonic personal aerosol sampler. *Indoor Air* 27(2): 409–416.

[17] Wang, Y., Wang, Y., Liu, X., Liu, W., Chen, D., Changju, W., Xie, J. (2019). An aerosol sensor for PM1 concentration detection based on 3D printed virtual impactor and SAW sensor. *Sensors and Actuators. A, Physical* 288, 67–74.

[18] Chiriacò, M.S., Rizzato, S., Primiceri, E., Spagnolo, S., Monteduro, A.G., Ferrara, F., Maruccio, G. (2018). Optimization of SAW and EIS sensors suitable for environmental particulate monitoring. *Microelectronic Engineering* 202:31–36.

[19] Liu, J., Hao, W., Liu, M., Liang, Y. and He, S. (2018). A novel particulate matter 2.5 sensor based on surface acoustic wave technology. *Applied Sciences* 8(1):82.

[20] Budde, M., El Masri, R., Riedel, T., Beigl, M. (2013). In *Enabling low-cost particulate matter measurement for participatory sensing scenarios*, Proceedings of the 12th international conference on mobile and ubiquitous multimedia, 2013; ACM; p 19.

[21] Snik, F., Rietjens, J.H., Apituley, A., Volten, H., Mijling, B., Di Noia, A., Heikamp, S., Heinsbroek, R.C., Hasekamp, O.P., Smit, J.M. (2014). Mapping atmospheric aerosols with a citizen science network of smartphone spectropolarimeters. *Geophysical Research Letters* 41 (20):7351–7358.

[22] Du, Z., Tsow, F., Wang, D., Tao, N. (2018). A miniaturized particulate matter sensing platform based on CMOS imager and real-time image processing. *IEEE Sensors Journal* 18(18): 7421–7428.

[23] Carminati, M., Ferrari, G., Sampietro, M. (2017). Emerging miniaturized technologies for airborne particulate matter pervasive monitoring. *Measurement* 101:250–256.

[24] Yang, J., Shen, R., Wang, C., Li, X., Liu, Y., Xu, K., Chen, W. (2018). In *A novel sensor for detecting PM 2.5 concentration based on refractive index sensing of a photonic crystal fiber*, Photonic and Phononic Properties of Engineered Nanostructures VIII, 2018; International Society for Optics and Photonics; p 105412K.

[25] Zhu, J., Özdemir, Ş.K., He, L., Chen, D.-R., Yang, L. (2011). Single virus and nanoparticle size spectrometry by whispering-gallery-mode microcavities. *Optics Express* 19(17):16195–16206.

[26] Toledo, J., Ruiz-Díez, V., Bertke, M., Suryo Wasisto, H., Peiner, E., Sánchez-Rojas, J.L. (2019). Piezoelectric MEMS resonators for cigarette particle detection. *Micromachines* 10(2):145.

[27] Paprotny, I., Doering, F., Solomon, P.A., White, R.M., Gundel, L.A. (2013). Microfabricated air-microfluidic sensor for personal monitoring of airborne particulate matter: Design, fabrication, and experimental results. *Sensors and Actuators. A, Physical* 201:506–516.

[28] Zhao, J., Liu, M., Liang, L., Wang, W., Xie, J. (2016). Airborne particulate matter classification and concentration detection based on 3D printed virtual impactor and quartz crystal microbalance sensor. *Sensors and Actuators. A, Physical* 238:379–388.

[29] Austin, E., Novosselov, I., Seto, E., Yost, M.G. (2015). Laboratory evaluation of the Shinyei PPD42NS low-cost particulate matter sensor. *PloS One* 10(9):e0137789.

[30] Wang, Y., Li, J., Jing, H., Zhang, Q., Jiang, J., Biswas, P. (2015). Laboratory evaluation and calibration of three low-cost particle sensors for particulate matter measurement. *Aerosol Science and Technology* 49(11):1063–1077.

[31] Liu, D., Zhang, Q., Jiang, J., Chen, D.-R. (2017). Performance calibration of low-cost and portable particular matter (PM) sensors. *Journal of Aerosol Science* 112:1–10.

[32] Hapidin, D.A., Saputra, C., Maulana, D.S., Munir, M.M., Khairurrijal, K. (2019). Aerosol chamber characterization for commercial particulate matter (PM) sensor evaluation. *Aerosol and Air Quality Research* 19(1):181–194.

[33] Sousan, S., Koehler, K., Thomas, G., Park, J.H., Hillman, M., Halterman, A., Peters, T.M. (2016). Inter-comparison of low-cost sensors for measuring the mass concentration of occupational aerosols. *Aerosol Science and Technology* 50(5):462–473.

[34] Kelly, K., Whitaker, J., Petty, A., Widmer, C., Dybwad, A., Sleeth, D., Martin, R., Butterfield, A. (2017). Ambient and laboratory evaluation of a low-cost particulate matter sensor. *Environmental Pollution* 221:491–500.

[35] Johnson, K.K., Bergin, M.H., Russell, A.G., Hagler, G.S. (2018). Field test of several low-cost particulate matter sensors in high and low concentration urban environments. *Aerosol and Air Quality Research* 18:565–578.
[36] Bewick, V., Cheek, L., Ball, J. (2003). Statistics review 7: Correlation and regression. *Critical Care* 7(6):451.
[37] Taylor, R. (1990). Interpretation of the correlation coefficient: A basic review. *Journal of Diagnostic Medical Sonography* 6(1):35–39.
[38] Sedgwick, P. (2012). Pearson's correlation coefficient. *British Medical Journal* 345:e4483.
[39] Smith, R.J. (2009). Use and misuse of the reduced major axis for line-fitting. *American Journal of Physical Anthropology: The Official Publication of the American Association of Physical Anthropologists* 140(3):476–486.
[40] Clarke, M. (1980). The reduced major axis of a bivariate sample. *Biometrika* 67(2):441–446.
[41] Ayers, G. (2001). Comment on regression analysis of air quality data. *Atmospheric Environment* 35(13):2423–2425.
[42] Gao, M., Cao, J., Seto, E. (2015). A distributed network of low-cost continuous reading sensors to measure spatiotemporal variations of PM2. 5 in Xi'an, China. *Environmental Pollution* 199:56–65.
[43] Weakliem, D.L. (1999). A critique of the Bayesian information criterion for model selection. *Sociological Methods & Research* 27(3):359–397.
[44] Watanabe, S. (2013). A widely applicable Bayesian information criterion. *Journal of Machine Learning Research* 14(Mar):867–897.
[45] Neath, A.A., Cavanaugh, J.E. (2012). The Bayesian information criterion: Background, derivation, and applications. *Wiley Interdisciplinary Reviews. Computational Statistics* 4(2): 199–203.
[46] Schwarz, G. (1978). Estimating the dimension of a model. *The Annals of Statistics* 6(2): 461–464.
[47] Vrieze, S.I. (2012). Model selection and psychological theory: A discussion of the differences between the Akaike information criterion (AIC) and the Bayesian information criterion (BIC). *Psychological Methods* 17(2):228.
[48] Zikova, N., Masiol, M., Chalupa, D., Rich, D., Ferro, A., Hopke, P. (2017). Estimating hourly concentrations of PM2. 5 across a metropolitan area using low-cost particle monitors. *Sensors* 17(8):1922.
[49] Wilcoxon, F. (1945). Individual comparisons by ranking methods. *Biometrics Bulletin* 1(6): 80–83.
[50] Whitley, E., Ball, J. (2002). Statistics review 6: Nonparametric methods. *Critical Care* 6 (6):509.
[51] Crilley, L.R., Shaw, M., Pound, R., Kramer, L.J., Price, R., Young, S., Lewis, A.C., Pope, F.D. (2018). Evaluation of a low-cost optical particle counter (Alphasense OPC-N2) for ambient air monitoring. *Atmospheric Measurement Techniques* 11(2): 709–720.
[52] Sousan, S., Gray, A., Zuidema, C., Stebounova, L., Thomas, G., Koehler, K., Peters, T. (2018). Sensor selection to improve estimates of particulate matter concentration from a low-cost network. *Sensors* 18(9):3008.
[53] Chen, -C.-C., Kuo, C.-T., Chen, S.-Y., Lin, C.-H., Chue, -J.-J., Hsieh, Y.-J., Cheng, C.-W., Wu, C.-M., Huang, C.-M. (2018). In *Calibration of Low-Cost Particle Sensors by Using Machine-Learning Method*, 2018 IEEE Asia Pacific Conference on Circuits and Systems (APCCAS), 2018; IEEE; pp 111–114.
[54] Li, J., Li, H., Ma, Y., Wang, Y., Abokifa, A.A., Lu, C., Biswas, P. (2018). Spatiotemporal distribution of indoor particulate matter concentration with a low-cost sensor network. *Building and Environment* 127:138–147.

[55] Zimmerman, N., Presto, A.A., Kumar, S.P., Gu, J., Hauryliuk, A., Robinson, E.S., Robinson, A.L., Subramanian, R. (2018). A machine learning calibration model using random forests to improve sensor performance for lower-cost air quality monitoring. *Atmospheric Measurement Techniques* 11:1.
[56] Li, C.-R., Lin, Y.-C., Hung, M.-W., Yang, -C.-C., Tsai, H.-Y., Chang, Y.-J., Huang, K.-C., Hsiao, W.-T. (2018). In *Integrating temperature, humidity, and optical aerosol sensors for a wireless module for three-dimensional space monitoring*, 2018 IEEE Sensors Applications Symposium (SAS), 2018; IEEE; pp 1–4.
[57] Wang, K., Chen, F.-E., Au, W., Zhao, Z., Xia, Z.-L. (2019). Evaluating the feasibility of a personal particle exposure monitor in outdoor and indoor microenvironments in Shanghai, China. *International Journal of Environmental Health Research* 29(2):209–220.
[58] Jayaratne, R., Liu, X., Thai, P., Dunbabin, M., Morawska, L. (2018). The influence of humidity on the performance of a low-cost air particle mass sensor and the effect of atmospheric fog. *Atmospheric Measurement Techniques* 11(8):4883–4890.
[59] Sayahi, T., Butterfield, A., Kelly, K. (2019). Long-term field evaluation of the Plantower PMS low-cost particulate matter sensors. *Environmental Pollution* 245:932–940.
[60] Zheng, T., Bergin, M.H., Johnson, K.K., Tripathi, S.N., Shirodkar, S., Landis, M.S., Sutaria, R., Carlson, D.E. (2018). Field evaluation of low-cost particulate matter sensors in high-and low-concentration environments. *Atmospheric Measurement Techniques* 11(8):4823–4846.
[61] Zamora, L.M., Xiong, F., Gentner, D., Kerkez, B., Kohrman-Glaser, J., Koehler, K. (2018). Field and laboratory evaluations of the low-cost plantower particulate matter sensor. *Environmental Science & Technology* 53(2):838–849.
[62] Di Antonio, A., Popoola, O., Ouyang, B., Saffell, J., Jones, R. (2018). Developing a relative humidity correction for low-cost sensors measuring ambient particulate matter. *Sensors* 18 (9):2790.
[63] N Genikomsakis, K., Galatoulas, N.-F., I Dallas, P., Candanedo Ibarra, L., Margaritis, D., S Ioakimidis, C. (2018). Development and on-field testing of low-cost portable system for monitoring PM2. 5 concentrations. *Sensors* 18(4):1056.
[64] Malings, C., Tanzer, R., Hauryliuk, A., Saha, P.K., Robinson, A.L., Presto, A.A., Subramanian, R. (2018). Fine particle mass monitoring with low-cost sensors: Corrections and long-term performance evaluation.
[65] Malings, C., Tanzer, R., Hauryliuk, A., Saha, P.K., Robinson, A.L., Subramanian, R., Presto, A.A. (2018). Correction and long-term performance evaluation of fine particulate mass monitoring with low-cost sensors.
[66] Waggoner, A.P., Weiss, R.E., Ahlquist, N.C., Covert, D.S., Will, S., Charlson, R.J. (1891–1909). Optical characteristics of atmospheric aerosols. *Atmospheric Environment (1967)* 1981(15):(10–11).
[67] McMurry, P.H., Stolzenburg, M.R. (1989). On the sensitivity of particle size to relative humidity for Los Angeles aerosols. *Atmospheric Environment (1967)* 23(2):497–507.
[68] Hess, M., Koepke, P., Schult, I. (1998). Optical properties of aerosols and clouds: The software package OPAC. *Bulletin of the American Meteorological Society* 79(5):831–844.
[69] Shettle, E.P., Fenn, R.W. (1979). Models for the aerosols of the lower atmosphere and the effects of humidity variations on their optical properties, Air Force Geophysics Lab Hanscom Afb Ma.
[70] Steele, H.M., Hamill, P. (1981). Effects of temperature and humidity on the growth and optical properties of sulphuric acid – Water droplets in the stratosphere. *Journal of Aerosol Science* 12(6):517–528.
[71] Kaiser, H., Specker, H. (1956). Bewertung und vergleich von analysenverfahren. *Fresenius' Journal of Analytical Chemistry* 149(1):46–66.

[72] EPA. (2016). 40 CFR Parts 58-Ambient Air Quality Surveillance (Subchapter C), Environmental Protection Agency, Washington, DC.
[73] Bartley, D.L., Shulman, S.A., Schlecht, P.C. (2003). Measurement uncertainty and NIOSH method accuracy range. *NIOSH Manual of Analytical Methods 4th edn. NIOSH Publication*, (2003–154), 208–28.
[74] Long, R., Beaver, M., Williams, R., Kronmiller, K., Garvey, S. (2014). Procedures and Concepts of EPA's Ongoing Sensor Evaluation Efforts, EM (Air Waste Manage. Assoc.), 8. https://cfpub.epa.gov/si/si_public_record_report.cfm?Lab=NERL&dirEntryId=290592.
[75] Manikonda, A., Zíková, N., Hopke, P.K., Ferro, A.R. (2016). Laboratory assessment of low-cost PM monitors. *Journal of Aerosol Science* 102:29–40.
[76] Reece, S., Williams, R., Colón, M., Southgate, D., Huertas, E., O'Shea, M., Iglesias, A., Sheridan, P. (2018). Spatial-temporal analysis of PM2. 5 and NO2 concentrations collected using low-cost sensors in Peñuelas, Puerto Rico. *Sensors* 18(12):4314.
[77] Williams, R., Kilaru, V., Snyder, E., Kaufman, A., Dye, T., Rutter, A., Russell, A., Hafner, H. (2014). Air sensor guidebook. US Environmental Protection Agency, Washington, DC. EPA/600/R-14/159 (NTIS PB2015-100610).
[78] Clements, A.L., Griswold, W.G., Rs, A., Johnston, J.E., Herting, M.M., Thorson, J., Collier-Oxandale, A., Hannigan, M. (2017). Low-cost air quality monitoring tools: From research to practice (a workshop summary). Sensors 17(11):2478.
[79] Papapostolou, V., Zhang, H., Feenstra, B.J., Polidori, A. (2017). Development of an environmental chamber for evaluating the performance of low-cost air quality sensors under controlled conditions. *Atmospheric Environment* 171:82–90.
[80] Li, H.Z., Gu, P., Ye, Q., Zimmerman, N., Robinson, E.S., Subramanian, R., Apte, J.S., Robinson, A.L., Presto, A.A. (2019). Spatially dense air pollutant sampling: Implications of spatial variability on the representativeness of stationary air pollutant monitors. *Atmospheric Environment: X* 2:100012. ISSN 2590–1621. https://doi.org/10.1016/j.aeaoa.2019.100012.
[81] Saha, P.K., Zimmerman, N., Malings, C., Hauryliuk, A., Li, Z., Snell, L., Subramanian, R., Lipsky, E., Apte, J.S., Robinson, A.L. (2019). Quantifying high-resolution spatial variations and local source impacts of urban ultrafine particle concentrations. *Science of the Total Environment* 655:473–481.
[82] Steinle, S., Reis, S. and Sabel, C.E. (2013). Quantifying human exposure to air pollution—Moving from static monitoring to spatio-temporally resolved personal exposure assessment. *Science of the Total Environment* 443:184–193.
[83] Ngo, N.S., Asseko, S.V.J., Ebanega, M.O., Allo'o, S.M.A.O., Hystad, P. (2018). The relationship among PM2. 5, traffic emissions, and socioeconomic status: Evidence from Gabon using low-cost, portable air quality monitors. *Transportation Research Part D: Transport and Environment* 68:2–9.
[84] Blanco, M.N., Fenske, R.A., Kasner, E.J., Yost, M.G., Seto, E., Austin, E. (2019). Real-time monitoring of spray drift from three different Orchard Sprayers. *Chemosphere* 222, 46–55.
[85] Schwarz, A.D., Meyer, J., Dittler, A. (9), 1826–18322018). Opportunities for low-cost particulate matter sensors in filter emission measurements. *Chemical Engineering & Technology* 41(9):1826–1832.
[86] Mazaheri, M., Clifford, S., Yeganeh, B., Viana, M., Rizza, V., Flament, R., Buonanno, G., Morawska, L. (2018). Investigations into factors affecting personal exposure to particles in urban microenvironments using low-cost sensors. *Environment International* 120:496–504.
[87] Subramanian, R., Ellis, A., Torres-Delgado, E., Tanzer, R., Malings, C., Rivera, F., Morales, M., Baumgardner, D., Presto, A., Mayol-Bracero, O.L. (2018). Air quality in puerto rico in the aftermath of hurricane maria: A case study on the use of lower cost air quality monitors. *ACS Earth and Space Chemistry* 2(11):1179–1186.

[88] Patel, S., Li, J., Pandey, A., Pervez, S., Chakrabarty, R.K., Biswas, P. (2017). Spatio-temporal measurement of indoor particulate matter concentrations using a wireless network of low-cost sensors in households using solid fuels. *Environmental Research* 152:59–65.
[89] Curto, A., Donaire-Gonzalez, D., Barrera-Gómez, J., Marshall, J.D., Nieuwenhuijsen, M.J., Wellenius, G.A., Tonne, C. (2018). Performance of low-cost monitors to assess household air pollution. *Environmental Research* 163:53–63.
[90] Dobson, R., Semple, S. (2018). How do you know those particles are from cigarettes?": An algorithm to help differentiate second-hand tobacco smoke from background sources of household fine particulate matter. *Environmental Research* 166:344–347.
[91] Thomas, G., Sousan, S., Tatum, M., Liu, X., Zuidema, C., Fitzpatrick, M., Koehler, K., Peters, T. (2018). Low-cost, distributed environmental monitors for factory worker health. *Sensors* 18 (5):1411.
[92] Cheng, H., Wang, L., Wang, D., Zhang, J., Cheng, L., Yao, P., Zhang, Z., Di Narzo, A., Shen, Y., Yu, J. (2019). Bio3Air, an integrative system for monitoring individual-level air pollutant exposure with high time and spatial resolution. *Ecotoxicology and Environmental Safety* 169: 756–763.
[93] Rogulski, M. (2018). Using low-cost PM monitors to detect local changes of air quality. *Polish Journal of Environmental Studies* 27:4.
[94] Penza, M., Suriano, D., Villani, M.G., Spinelle, L., Gerboles, M.. (2014). In *Towards air quality indices in smart cities by calibrated low-cost sensors applied to networks*, SENSORS, 2014 IEEE, 2014; IEEE; pp 2012–2017.
[95] Steinle, S., Reis, S., Sabel, C.E., Semple, S., Twigg, M.M., Braban, C.F., Leeson, S.R., Heal, M.R., Harrison, D., Lin, C. (2015). Personal exposure monitoring of PM2. 5 in indoor and outdoor microenvironments. *Science of the Total Environment* 508:383–394.
[96] Leaffer, D., Wolfe, C., Doroff, S., Gute, D., Wang, G., Ryan, P. (2019). Wearable ultrafine particle and noise monitoring sensors jointly measure personal co-exposures in a pediatric population. *International Journal of Environmental Research and Public Health* 16(3):308.
[97] Hu, R., Wang, S., Aunan, K., Zhao, M., Chen, L., Liu, Z., Hansen, M.H. (2019). Personal exposure to PM 2.5 in Chinese rural households in the Yangtze River Delta. *Indoor Air* 29(3):403–412.
[98] SM, S.N., Yasa, P.R., Narayana, M., Khadirnaikar, S., Rani, P. (2019). Mobile monitoring of air pollution using low cost sensors to visualize spatio-temporal variation of pollutants at urban hotspots. *Sustainable Cities and Society* 44:520–535.
[99] Yang, F., Lau, C.F., Tong, V.W.T., Zhang, K.K., Westerdahl, D., Ng, S., Ning, Z. (2019). Assessment of personal integrated exposure to fine particulate matter of urban residents in Hong Kong. *Journal of the Air & Waste Management Association* 69(1):47–57.
[100] Cheng, K.-C., Tseng, C.-H., Hildemann, L.M. (2019). Using indoor positioning and mobile sensing for spatial exposure and environmental characterizations: Pilot demonstration of PM2. 5 mapping. *Environmental Science & Technology Letters* 6(3):153–158.
[101] Gu, Q., Jia, C. (2019). In *A Consumer UAV-based Air Quality Monitoring System for Smart Cities*, 2019 IEEE International Conference on Consumer Electronics (ICCE), 2019; IEEE; pp 1–6.
[102] Mayuga, G.P., Favila, C., Oppus, C., Macatulad, E., Lim, L.H. (2018). In *Airborne Particulate Matter Monitoring Using UAVs for Smart Cities and Urban Areas*, TENCON 2018–2018 IEEE Region 10 Conference, 2018; IEEE; pp 1398–1402.
[103] Cashikar, A., Li, J., Biswas, P. (2019). Particulate matter sensors mounted on a robot for environmental aerosol measurements. *Journal of Environmental Engineering* 145 (10):04019057.

[104] Wilson, J.G., Kingham, S., Pearce, J., Sturman, A.P. (2005). A review of intraurban variations in particulate air pollution: Implications for epidemiological research. *Atmospheric Environment* 39(34):6444–6462.

[105] Wongphatarakul, V., Friedlander, S., Pinto, J. (1998). A comparative study of PM2. 5 ambient aerosol chemical databases. *Environmental Science & Technology* 32(24):3926–3934.

[106] Rajasegarar, S., Zhang, P., Zhou, Y., Karunasekera, S., Leckie, C., Palaniswami, M. (2014). In *High resolution spatio-temporal monitoring of air pollutants using wireless sensor networks*, 2014 IEEE Ninth International Conference on Intelligent Sensors, Sensor Networks and Information Processing (ISSNIP), 2014; IEEE; pp 1–6.

[107] Rajasegarar, S., Havens, T.C., Karunasekera, S., Leckie, C., Bezdek, J.C., Jamriska, M., Gunatilaka, A., Skvortsov, A., Palaniswami, M. (2014). High-resolution monitoring of atmospheric pollutants using a system of low-cost sensors. *IEEE Transactions on Geoscience and Remote Sensing* 52(7):3823–3832.

[108] Lam, -N.S.-N. (1983). Spatial interpolation methods: A review. *The American Cartographer* 10 (2):129–150.

[109] Li, J., Heap, A.D. (2008). A review of spatial interpolation methods for environmental scientists.

[110] Li, J., Heap, A.D. (2011). A review of comparative studies of spatial interpolation methods in environmental sciences: Performance and impact factors. *Ecological Informatics* 6 (3–4):228–241.

[111] Mitas, L., Mitasova, H. (1999). Spatial interpolation. In: P.Longley, M.F. Goodchild, D.J. Maguire, D.W.Rhind (Eds.), Geographical Information Systems: Principles, Techniques, Management and Applications, GeoInformation International, Wiley, 481–492.

[112] Masiol, M., Zíková, N., Chalupa, D.C., Rich, D.Q., Ferro, A.R., Hopke, P.K. (2018). Hourly land-use regression models based on low-cost PM monitor data. *Environmental Research* 167: 7–14.

[113] Kelly, K.E., Xing, W.W., Sayahi, T., Mitchell, L., Becnel, T., Gaillardon, P.-E., Meyer, M., Whitaker, R.T. (2020). Community-based measurements reveal unseen differences during air pollution episodes. *Environmental Science & Technology* 55(1):120–128.

[114] Eilenberg, S.R., Subramanian, R., Malings, C., Hauryliuk, A., Presto, A.A., Robinson, A.L. (2020). Using a network of lower-cost monitors to identify the influence of modifiable factors driving spatial patterns in fine particulate matter concentrations in an urban environment. *Journal of Exposure Science & Environmental Epidemiology* 30(6):949–961.

[115] Tanzer, R., Malings, C., Hauryliuk, A., Subramanian, R., Presto, A.A. (2019). Demonstration of a low-cost multi-pollutant network to quantify intra-urban spatial variations in air pollutant source impacts and to evaluate environmental justice. *International Journal of Environmental Research and Public Health* 16(14):2523.

[116] Tanzer-Gruener, R., Li, J., Eilenberg, S.R., Robinson, A.L., Presto, A.A. (2020). Impacts of modifiable factors on ambient air pollution: A case study of COVID-19 shutdowns. *Environmental Science & Technology Letters* 7(8):554–559.

[117] Gupta, P., Doraiswamy, P., Levy, R., Pikelnaya, O., Maibach, J., Feenstra, B., Polidori, A., Kiros, F., Mills, K. (2018). Impact of California fires on local and regional air quality: The role of a low-cost sensor network and satellite observations. *GeoHealth* 2(6):172–181.

[118] Li, J., Zhang, H., Chao, C.-Y., Chien, C.-H., Wu, C.-Y., Luo, C.H., Chen, L.-J., Biswas, P. (2020). Integrating low-cost air quality sensor networks with fixed and satellite monitoring systems to study ground-level PM2. 5. *Atmospheric Environment* 223:117293.

[119] Malings, C., Westervelt, D.M., Hauryliuk, A., Presto, A.A., Grieshop, A., Bittner, A., Beekmann, M., Subramanian, R. (2020). Application of low-cost fine particulate mass

monitors to convert satellite aerosol optical depth to surface concentrations in North America and Africa. *Atmospheric Measurement Techniques* 13(7):3873–3892.
[120] Feinberg, S., Williams, R., Hagler, G.S., Rickard, J., Brown, R., Garver, D., ... Garvey, S. (2018). Long-term evaluation of air sensor technology under ambient conditions in Denver, Colorado. *Atmospheric Measurement Techniques* 11(8):4605–4615.
[121] Sousan, S., Koehler, K., Hallett, L., Peters, T.M. (2016). Evaluation of the Alphasense optical particle counter (OPC-N2) and the Grimm portable aerosol spectrometer (PAS-1.108). *Aerosol Science and Technology* 50(12):1352–1365.
[122] Jiao, W., Hagler, G., Williams, R., Sharpe, R., Brown, R., Garver, D., ... Buckley, K. (2016). Community Air Sensor Network (CAIRSENSE) project: evaluation of low-cost sensor performance in a suburban environment in the southeastern United States. *Atmospheric Measurement Techniques* 9(11):5281–5292
[123] Mukherjee, A., Stanton, L.G., Graham, A.R., Roberts, P.T. (2017). Assessing the utility of low-cost particulate matter sensors over a 12-week period in the Cuyama valley of California. *Sensors* 17(8):1805.
[124] Gillooly, S.E., Zhou, Y., Vallarino, J., Chu, M.T., Michanowicz, D.R., Levy, J.I., Adamkiewicz, G. (2019). Development of an in-home, real-time air pollutant sensor platform and implications for community use. *Environmental Pollution* 244:440–450.
[125] Semple, S., Ibrahim, A.E., Apsley, A., Steiner, M., Turner, S. (2015). Using a new, low-cost air quality sensor to quantify second-hand smoke (SHS) levels in homes. *Tobacco control* 24(2):153–158.
[126] Semple, S., Apsley, A., MacCalman, L. (2013). An inexpensive particle monitor for smoker behaviour modification in homes. *Tobacco Control* 22(5):295–298.
[127] Jovašević-Stojanović, M., Bartonova, A., Topalović, D., Lazović, I., Pokrić, B., Ristovski, Z. (2015). On the use of small and cheaper sensors and devices for indicative citizen-based monitoring of respirable particulate matter. Environmental Pollution 206:696–704.
[128] Han, I., Symanski, E., Stock, T.H. (2017). Feasibility of using low-cost portable particle monitors for measurement of fine and coarse particulate matter in urban ambient air. *Journal of the Air & Waste Management Association* 67(3):330–340.
[129] Franken, R., Maggos, T., Stamatelopoulou, A., Loh, M., Kuijpers, E., Bartzis, J., ... Pronk, A. (2019). Comparison of methods for converting Dylos particle number concentrations to PM2. 5 mass concentrations. *Indoor Air* 29(3):450–459.
[130] Moreno-Rangel, A., Sharpe, T., Musau, F., McGill, G. (2018). Field evaluation of a low-cost indoor air quality monitor to quantify exposure to pollutants in residential environments. *Journal of Sensors and Sensor Systems* 7(1):373–388.

Liang-Yi Lin

Chapter 5 Carbon dioxide conversion methodologies

Abstract: Reduction of carbon dioxide (CO_2) emission has been regarded as a crucial step toward a sustainable environment. Transforming CO_2 waste into potentially clean fuels such as methane can be a permanent solution that mitigates the CO_2 emissions and achieves a sustainable energy future simultaneously. Particularly, the transformation of CO_2 via catalytic processes using the solar energy has attracted great interest because the abundant and inexpensive solar energy can be directly utilized. However, the conversion efficiencies of the existing photochemical systems are still far from satisfactory for commercialization. Integrating the photochemical processes with electro- or thermochemical processes for transforming CO_2 into carbon-containing fuels has attracted great attention, because this approach synergistically drives catalytic reactions and leads to greatly enhanced catalytic performances. This chapter summarizes the recent literature on (i) photochemical, (ii) photoelectrochemical, and (iii) photothermal processes employed for CO_2 conversion, and reaction mechanisms and the requirements for designing catalytically active materials to achieve high CO_2 conversion efficiency are discussed. Further, the recent progress on the aerosol processing of novel catalytic materials for CO_2 conversion applications is also briefly reviewed.

Keywords: CO_2 capture and conversion, nanocatalyst, photocatalysis, photothermal, photoelectrochemical, aerosol synthesis

5.1 Introduction

Global warming, an urgent environmental concern, is driven by the steady growth in fossil fuel combustion, significantly increasing the CO_2 concentrations in the atmosphere [1]. Carbon capture and sequestration (CCS), is recognized by the Intergovernmental Panel on Climate Change (IPCC) as one of the feasible options for CO_2 mitigation [2]. At present, capturing CO_2 from exhaust flue gases of stationary sources (e.g., coal-fired power plants) by liquid absorption processes is the most mature approach [3, 4]. However, stripping the CO_2 process incurs an energy penalty [5]. Furthermore, finding the appropriate storage sites is difficult because of environmental concerns about storing CO_2 in the deep sea or underground [6]. Thus, a more viable process would reutilize the captured CO_2.

Liang-Yi Lin, Institute of Environmental Engineering, National Yang Ming Chiao Tung University, Hsinchu 300, Taiwan, e-mail: lylin@nctu.edu.tw

https://doi.org/10.1515/9783110729481-005

Technologies that convert CO_2 to synthetic fuels offer an alternative that simultaneously addresses both the environmental and energy crisis [1, 7]. Nevertheless, the transformation of CO_2 into fuels via chemical activation/reduction is scientifically challenging, because the extremely stable CO_2 molecule requires appropriate catalysts and significant energy input. Solar energy, an abundant resource, could inexpensively meet this need. Currently, technologies in utilization and chemical conversion of CO_2 are not mature and efficient enough for practical implementation and commercialization in terms of technology and economy [8, 9]. The major challenges include low conversion rates of CO_2, lack of catalytic stability in long-term operation, and insufficient product selectivity [10]. While considerable attention has been paid to developing efficient photochemical and photoelectrochemical systems to overcome these challenges over the past decades [10–12], recent works on catalytic transformation of CO_2 through the photothermal processes has achieved significant progresses [13, 14]. However, fundamental challenges related to photo(electro)chemistry, semiconductor physics and engineering, and heterogeneous catalysis still need be addressed [15–17].

Catalysis technology significantly depends on the catalyst selection. Considering the catalyst preparation methods, the review will pay attention to both catalytic processes and methods of preparation of corresponding catalysts. Potentially, technologies of catalyst preparation via gas-phase (aerosol) technology are more advantageous than other methods because of possible fine-control of active particle sizes and distributions. However, presently this aerosol technology for catalyst preparation is at the stage of development. This chapter provides an overview of the significant work in solar-driven catalytic CO_2 conversion. Moreover, the principles of each catalytic process are systematically discussed. Finally, this chapter is ended with the perspectives on future directions and urgent challenges for catalytic materials development for CO_2 conversion.

5.2 Photochemical reduction of CO_2

In recent decades, considerable attention has been devoted to solar-activated photochemical processes that convert CO_2 molecules into useful chemicals at ambient temperature and atmospheric pressure [18]. Photochemical processes are attractive because they directly utilize inexpensive and abundant solar energy. Various semiconductors such as ZnO, TiO_2, WO_3, and In_2O_3 have been ongoing subjects of study [19–23]. The complicated steps of photocatalysis are illustrated in Figure 5.1 [24]. When a semiconductor photocatalyst is illuminated by light with energy (E_{hv}) equal to or higher than its intrinsic bandgap (E_g), electrons can be excited from its valence band (VB) to a conduction band (CB), leaving a positively charged electron hole in the VB. These charge carriers are then separated from each other and initiate the photoreduction and photooxidation, respectively.

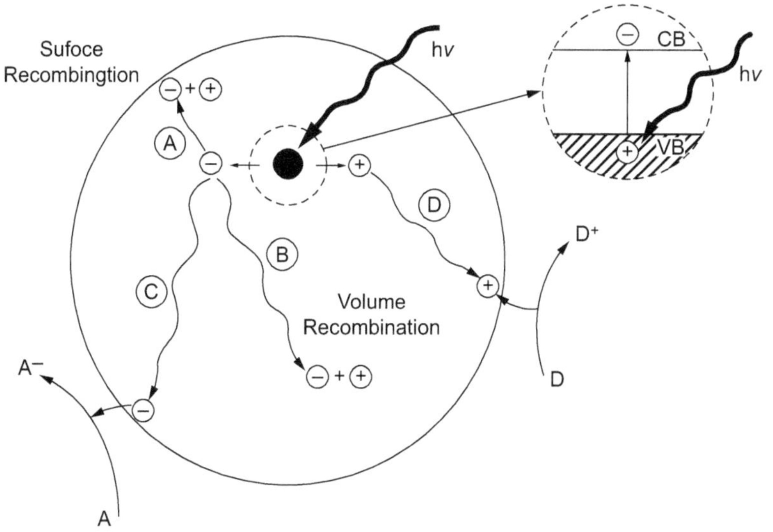

Figure 5.1: Schematic photoexcitation in a solid followed by deexcitation events [24].

The charge transfer kinetics between carriers on the catalyst surface is determined by the degree of correlation of the energy levels of the semiconductors and the redox agents in the solution at a predetermined pH value. Figure 5.2 displays

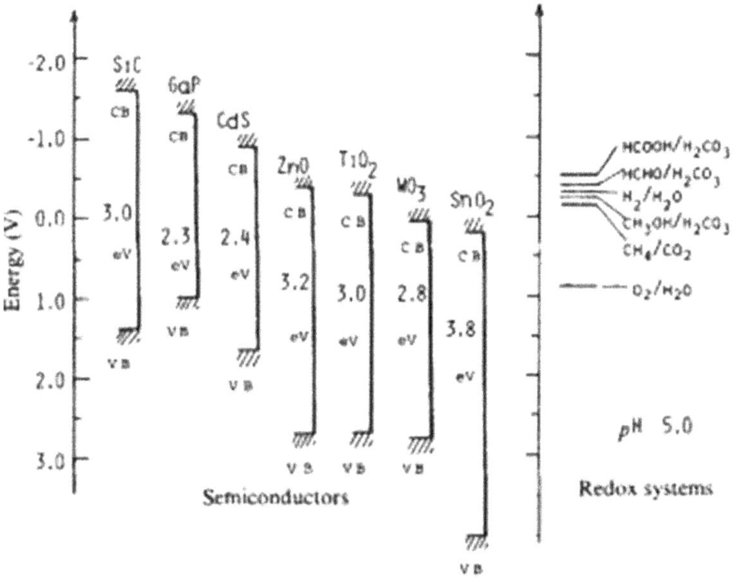

Figure 5.2: The energy correlation between semiconductors and redox agents in water [25].

the energy levels of various semiconductors, and their corresponding redox potentials in a CO_2 photoreduction system. In a water-assisted catalytic CO_2 reduction system, the oxidation potential of the holes is required to be strong enough to proceed the water oxidation to produce protons and O_2, i.e., for efficient photolysis the VB edge should be more positive than the energy level of water oxidation. In the meantime, the energy level of the electrons in the catalyst should be higher than the corresponding CO_2 reduction potential, which means that the CB potential of the catalyst should be more negative than the reduction potential for a certain product.

A proton-assisted multi-electron transfer pathway has been proposed for photo-reducing CO_2. Accordingly, the formation of carbon-containing products depends on the numbers of available electrons and protons in the chemical reaction, as illustrated in the eq. (5.1)–(5.8) [26]. For example, for CH_4 formation, eight electrons and protons are required, whereas the generation of CH_3OH needs six electrons and protons. Meanwhile, the formation of H_2 also compete the available electrons and protons (eq. (5.8) **in Scheme 1**). Suppressing such undesired competing reactions is thus of importance.

Reaction	Eo (V vs NHE)
$CO_2 + 2e^- \rightarrow CO^{2-}$ (5.1)	−1.9
$CO_2 + 2H^+ + 2e^- \rightarrow HCOOH$ (5.2)	−0.61
$CO_2 + 2H^+ + 2e^- \rightarrow CO + H_2O$ (5.3)	−0.53
$CO_2 + 4H^+ + 4e^- \rightarrow HCHO + H_2O$ (5.4)	−0.48
$CO_2 + 6H^+ + 6e^- \rightarrow CH_3OH + H_2O$ (5.5)	−0.38
$CO_2 + 8H^+ + 8e^- \rightarrow CH_4 + H_2O$ (5.6)	−0.24
$2H_2O + 4h^+ \rightarrow O_2 + 4H^+$ (5.7)	+0.81
$2H^+ + 2e^- \rightarrow H_2$ (5.8)	−0.42

Pan et al. [27] reported the photocatalytic CO_2 conversion by carbon-coated In_2O_3 nanobelts, using triethanolamine as a sacrificial reagent. Using Pt as a co-catalyst, they achieved CO and CH_4 yields as high as 126.6 and 27.9 μmol h^{-1}, respectively. In a promising one-step system for carbon capture and utilization, Huang et al. [28, 29] replaced water with an ammonia solution and achieved a high CO production rate of 550.7 μmol h^{-1}. Using H_2O as an electron source inherently lowers the activity and selectivity in photocatalytic conversion of CO_2. Without a sacrificial reagent, the oxidation of H_2O competes with the reduction of CO_2. However, in practice, water remains the primary reactant because of its readily availability and low cost.

Although photocatalysts for CO_2 conversion still face such challenges as the low catalytic activity and insufficient stability, they have attracted great attention. Particularly, transition metal oxide-based semiconductors have been frequently studied [30, 31]. In a typical semiconductor-based photocatalytic system, the type of photocatalyst plays a crucial role in governing the photocatalytic activity and reaction kinetics. The catalytic activity of semiconductor photocatalysts is strongly influenced by several parameters, including the bandgap of the semiconductor, the separation efficiency of charge carriers, and the number and availability of active sites. As one of the most studied photocatalysts, TiO_2 has received considerable attention due to its thermal/mechanical stability, high redox potential, and nontoxicity [32]. However, its insufficient photocatalytic performance has so far made its use impractical. It absorbs only UV light, and the photogenerated charge carriers recombine rapidly. Moreover, catalytic decay is another major concern with TiO_2-based materials, associated with the loss of available reaction sites during prolonged reaction.

Numerous strategies to modify TiO_2 have been demonstrated, such as heterojunction construction [33–36], impurity doping [37–40], defect engineering [41–43] and metal deposition [44–48], inducing the visible sensitivity and suppressing the recombination rates of the charges [49]. For example, the deposition of Pt and Rh nanoparticles (NPs) as electron sinks has implications for the multi-electron transfer reactions, as illustrated in Figure 5.3. Xie and co-workers [50] performed the preparation of $MgO/Pt\text{-}TiO_2$ nanocomposites, and showed that the photoexcited electrons on TiO_2 are easily captured by the Pt NPs. The electron-rich Pt NPs thus favored the formation of CH_4, which is more thermodynamically feasible than the formation of CO. Deposition with Au [51, 52] or Ag [53, 54] can also extend the light response to the visible spectrum because of their strong localized surface plasmonic resonance properties.

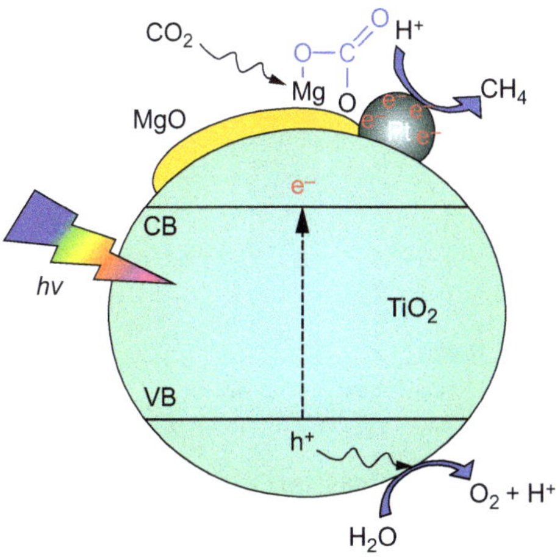

Figure 5.3: Reaction mechanisms of H_2O-assisted photoreduction of CO_2 over the MgO layer and Pt nanoparticles on TiO_2 [50].

Graphene-based photocatalytic heterostructures are promising materials in a variety of energy-related applications [55–58]. The morphological variety and electronic versatility make graphene a promising candidate for producing solar fuels, such as photocatalytically splitting water to make H_2, and photocatalytically reducing CO_2 to make hydrocarbons [59, 60]. Coupling graphene derivatives with various semiconductor photocatalysts has demonstrated to greatly suppress the photogenerated carrier recombination [61, 62], increased CO_2 adsorption capacity [63, 64] and improve the photostability [63] and the light absorption [65–67] of the composite photocatalyst, all of which together lead to improve the catalytic performances.

Yao et al. [68] synthesized a series of amine-coated crumpled reduced graphene oxide (rGO)/TiO_2 composites that, compared to the parent TiO_2, displayed superior photocatalytic activity and stability in reducing CO_2 into CO. Cho and co-workers demonstrated that amine-modified rGO/CdS performed remarkably elevated visible-light photoactivity in reducing CO_2 to CH_4, compared to bare CdS and rGO/CdS (Figure 5.4) [69]. In addition to graphene oxide's co-catalytic activity, some reports have also explored its intrinsic activity. As reported by Hsu et al. [70], graphene oxide can reduce CO_2 into CH_3OH by H_2O. Moreover, depositing metallic Cu particles on GO significantly favors the production of CH_3OH over that of blank GO, from 0.172 μmol g cat^{-1} h^{-1} to 6.4 μmol g cat^{-1} h^{-1}. This increase is also a 223-fold enhancement over the benchmark TiO_2-P25. Although a deeper insight into the promotional effects in different reaction systems remains to be revealed, all of these findings suggest that graphene is a highly potential cocatalyst/photocatalyst for enhancing the photochemical CO_2 conversion.

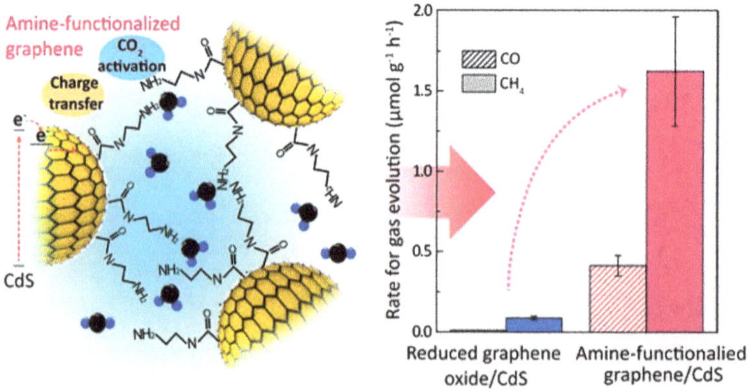

Figure 5.4: **(A)** The photoreduction of CO_2 on amine-modified rGO/CdS, **(B)** Comparison of CO/CH_4 ratio on rGO/CdS and amine-modified rGO/CdS [69].

A family of metal-organic-frameworks (MOFs) is another class of novel materials for CO_2 capture and photoreduction [71–78]. He and co-workers [79] prepared a TiO_2/HKUST-1 ((Cu-benzene-1,3,5-tricarboxylate) composite via an aerosol-assisted spray technique. During the aerosol process, the HKUST-1 crystals were formed via the self-assembly and the TiO_2 nanoparticles were uniformly distributed within the matrix of HKUST-1 (Figure 5.5A and B). While the crystallinity of the MOFs was not altered in the presence of TiO_2, the resulting composites showed much-enhanced adsorption toward both CO_2 and H_2O, which is beneficial for enhancing both the reaction rate (Figure 5.5D) and the production yield (Figure 5.5C). Similarly, Li et al. [80] created a TiO_2/HKUST-1 composite, which contained around 33 wt.% TiO_2. The TiO_2/HKUST-1 showed a fivefold higher CH_4 production per gram of TiO_2 compared to that of the pristine TiO_2. Both effective charge separation and high electron density together contributed to the increased CH_4 evolution. The molecular simulations suggested that adding extra electrons to the Cu centers facilitated the adsorption and activation of CO_2, whereas the transient absorption spectroscopy further verified the electron transfer from the CB of TiO_2 to the Cu sites on the HKUST-1. Overall, the Cu sites on HKUST-1 were capable of adsorbing the CO_2 molecules, and would become catalytically active when received the electrons from the neighboring TiO_2. This demonstration clearly reveals a synergistic effect between the CO_2 adsorbents and photocatalysts that lead to improve the catalytic performances.

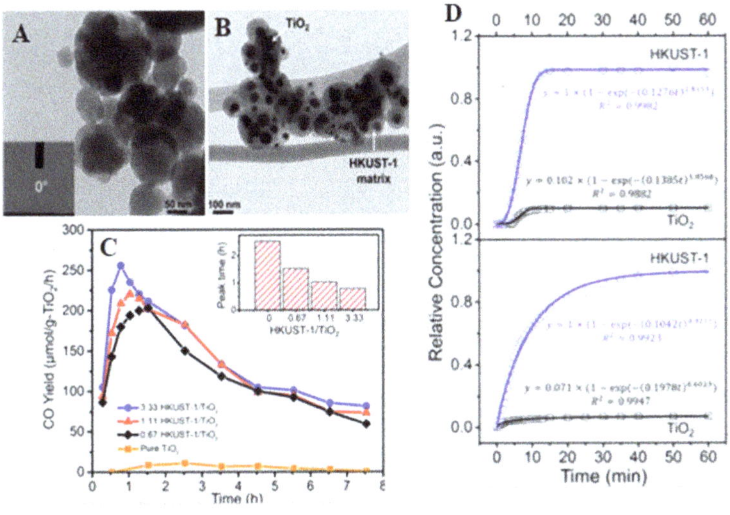

Figure 5.5: TEM images of aerosol-made **(A)** HKUST-1 and **(B)** TiO_2/HKUST-1; **(C)** The CO production yields on TiO_2 and TiO_2/HKUST-1; **(D)** Adsorption measurements of CO_2 and H_2O [79].

5.3 Photoelectrochemical reduction of CO_2

Photoelectrochemical CO_2 reduction is another method for producing solar fuels [81–83]. In this catalytic system, the recombination of photoexcited charge carriers can be hindered in the presence of an additional electric field at the semiconductor-electrolyte interface, and thus the lifetime of the photoexcited charge carriers is effectively prolonged [84]. Hence, combining photochemical and electrochemical reduction for CO_2 conversion provides the benefits of both approaches, and leads to an highly efficient and selective conversion of CO_2. The photoelectrochemical reduction of CO_2 in H_2O, using a photocathode in combination with a photoanode, is a useful strategy for reducing CO_2 without an externally applied electrical bias. To ensure the electron transfer from the photoanode to the photocathode through the external wire, the CB of the photoanode for H_2O oxidation should be above the VB edge of the photocathode for CO_2 reduction, (Figure 5.6).

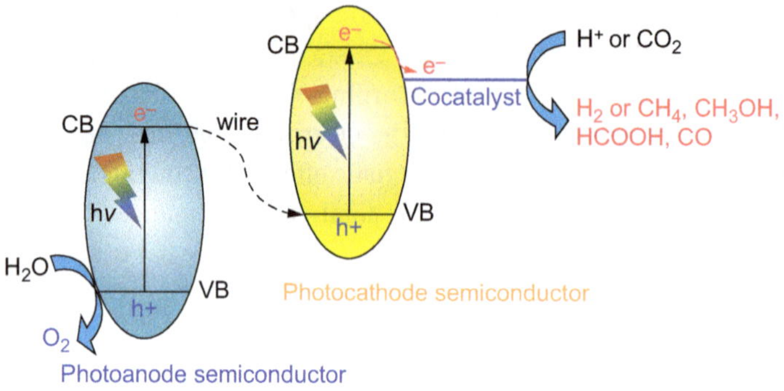

Figure 5.6: Scheme of a full photoelectrocatalytic cell integrating CO_2 reduction and H_2O oxidation [10].

Sato et al. [85] carried out the photoelectrocatalytic conversion of CO_2 to $HCOO^-$ on an InP/Ru complex polymer hybrid. By coupling the hybrid photocatalyst with TiO_2 for H_2O oxidation, they achieved the selective CO_2 photoreduction to $HCOO^-$ in an aqueous medium, where H_2O was used as both a proton source and an electron donor. The so-called Z-scheme (Figure 5.7) system operated with no external electrical bias. The conversion efficiency of solar energy to chemical energy was 0.03–0.04%, whereas the selectivity of $HCOO^-$formation was higher than 70%. In other research on photo-electrochemical reduction of CO_2, Yu et al. [86] employed the atomic layer deposition method to grow a TiO_2 film on a black silicon (b-Si) photoanode. The optimized thin film outperformed the planar $Si/Co(OH)_2$ and $b\text{-}Si/Co(OH)_2$ counterparts, and achieved a photocurrent density of 32.3 mA cm^{-2} at an external potential of 1.48 V in 1 M NaOH electrolyte. Both the catalytic performance and stability of b-Si-based electrochemical systems were effectively improved by the ALD-made TiO_2 protection layer, and this hybrid system may open a new avenue for effective generation of solar fuels.

Even though the photoelectrochemical processes for CO_2 conversion hold great potential, significant technological advances are still needed to further improve the reaction rates and energy efficiency. In this regard, CO_2 adsorption is an essential step during the chemical conversion of CO_2, and the reaction kinetics are highly correlated

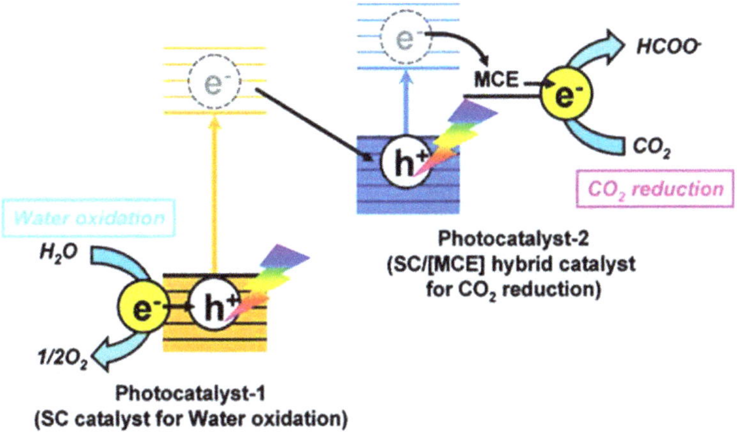

Figure 5.7: Total reaction of the Z-scheme system for CO_2 reduction [85].

with CO_2 concentration. Due to the low solubility of CO_2 in water (0.033 mol/L), its concentration on the catalyst surface is rather low. This slows the reaction kinetics, and severely deteriorates solid/liquid heterogeneous reduction. Recent efforts have been made to integrate inorganic semiconductors with molecular catalysts, such as Ru(II) or Re(II) bipyridyl complexes, to increase the fixation and activation of CO_2 [87]. Sahara et al. [88] developed a photocathode of a Ru(II) – Re(I) complex photocatalyst, and a CoO_x/TaON photoanode to realize a visible light-driven ($\lambda > 400$ nm) CO_2 conversion in a $NaHCO_3$ solution (pH = 8.3) with high efficiency and preferential production of CO (Figure 5.8). This is the first demonstration of a molecular/semiconductor hybrid system that achieved the CO_2 reduction under visible light illumination.

In addition to molecular catalysts, MOFs have also been recognized as promising candidates to facilitate selective adsorption and fixation of CO_2, enhance surface CO_2 concentrations, and lower the activation energy of the reaction to achieve efficient CO_2 reduction [89–93]. Shen et al. [94] demonstrated the fabrication of zeolite imidazolate framework (ZIF9) decorated Co_3O_4 nanowires (Figure 5.9), and the resulting thin film performed better than its counterpart Co_3O_4 nanowires and ZIF9. The improved activity resulted from synergy; there was more efficient charge transfer between the Co_3O_4 nanowires and ZIF9, and the ZIF9 increased CO_2 adsorption.

As far, studies on the photoelectrochemical CO_2 reduction over various catalysts have been intensively demonstrated; however, only a limited number of studies have investigated the CO_2 reduction kinetics [82, 95]. Peng et al. [96] conducted a kinetic study on the photoelectrochemical reduction of CO_2 based on the reduction mechanism: $CO_2 \rightarrow HCOOH \rightarrow HCHO \rightarrow CH_3OH \rightarrow CH_4$ (eq. (9–16)).

$$CO_2 + H. \rightarrow HCO_2 \tag{5.9}$$

$$CHO_2. + H. \rightarrow HCOOH \tag{5.10}$$

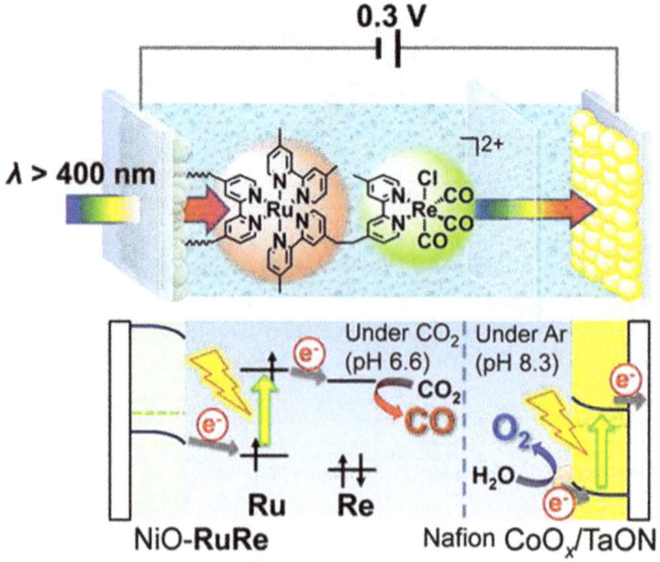

Figure 5.8: Schematic image of photoelectrochemical cell [88].

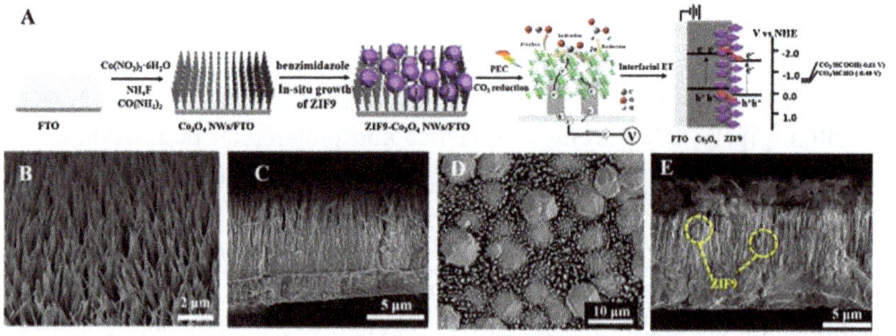

Figure 5.9: **(A)** Fabrication of the ZIF9-Co_3O_4 hybrid and reaction mechanism of PEC CO_2 reduction on ZIF9-Co_3O_4. **(B)** SEM image of top view of the Co_3O_4 nanowires. **(C)** SEM image of side view of Co_3O_4 nanowires. **(D)** SEM image of top view of the ZIF9-Co_3O_4. **(E)** SEM image of side view of ZIF9-Co_3O_4 [94].

$$HCOOH + H. \rightarrow CH_3O_2. \tag{5.11}$$

$$CH_{3.}O_2 + H. \rightarrow CH_2O + H_2O \tag{5.12}$$

$$CH_{2.}O + H. \rightarrow CH_3O. \tag{5.13}$$

$$CH_{3.} + H. \rightarrow CH_3OH \tag{5.14}$$

$$CH_{3.}OH + H. \rightarrow CH_{3.} + H_2O \tag{5.15}$$

$$CH_{3.} + H. \rightarrow CH_4 \tag{5.16}$$

Based on the above reaction schemes, the CO_2 reduction process are described by the following first-order reactions (eq. (17–21)),

$$\frac{d[CO_2]}{dt} = -K_1[CO_2] \tag{5.17}$$

$$\frac{d[HCOOH]}{dt} = -K_1[CO_2] - K_2[HCOOH\ ] \tag{5.18}$$

$$\frac{d[HCOH]}{dt} = -K_2[HCOOH] - K_3[HCOH] \tag{5.19}$$

$$\frac{d[CH_3OH]}{dt} = -K_3[HCOH] - K_4[CH_3OH] \tag{5.20}$$

$$\frac{d[CH_4]}{dt} = K_4[CH_3OH] \tag{5.21}$$

where t presents the reaction time and k_1, k_2, k_3, and k_4 are the reaction rate constants of CO_2, HCOOH, HCOH, and CH_3OH, respectively. A good agreement between the prediction and measurement was demonstrated with values of index of agreement > 0.87. Also, the rate constants were calculated as 6.68×10^{-7}, 2.18×10^{-4}, 8.62×10^{-4} and 2.27×10^{-4} (s^{-1}), for k_1, k_2, k_3, and k_4, respectively. One can note that k_1 is thousand times smaller than the others, probably because of the relatively high initial CO_2 concentration (0.13 M) in the cathodic chamber.

5.4 Photothermal reduction of CO_2

Photothermal catalysis, integrating both thermochemical and photochemical processes, has been proved to be a particularly effective approach for producing sustainable chemicals from CO_2 [97, 98]. Compared with the purely semiconductor-based photocatalytic systems where the catalytic performances are strongly restricted by their intrinsic bandgaps and redox potentials, one key advantage of the photothermal strategy is its effective harvesting of the whole solar spectrum that allows a greater utilization of the solar energy [99]. Also, the need of solar concentrators for producing high temperature to carry out the thermochemical process can be avoided [100].

Meng and co-workers [101] first explored the H_2-assisted photothermal CO_2 reduction into CH_4 over several Al_2O_3-supported Group VIII metals (Co, Rh, Ir, Ni, Fe, Pt, and Pd). Among all the catalysts studied, Al_2O_3 supported Ru and Rh exhibited high reaction rates, of 18.16 and 6.36 mol h^{-1} g^{-1} respectively, for CH_4 production. These production rates are several orders of magnitude higher than the corresponding photocatalytic CO_2-conversion rates (μmol h^{-1} g^{-1}). Wide-band-gap semiconductors ($E_g > 3$ eV), like ZnO, require UV irradiation as the excitation source, but photothermal CO_2 conversion can still proceed on Group VIII metals without UV light irradiation. Thus, these metals are intriguing for their efficient usage of solar energy, particularly the low-energy photons of visible and infrared light. Meng et al. also found that the photothermal effect induced a rapid increase in reaction temperature over these Group VIII–based catalysts (Figure 5.10).

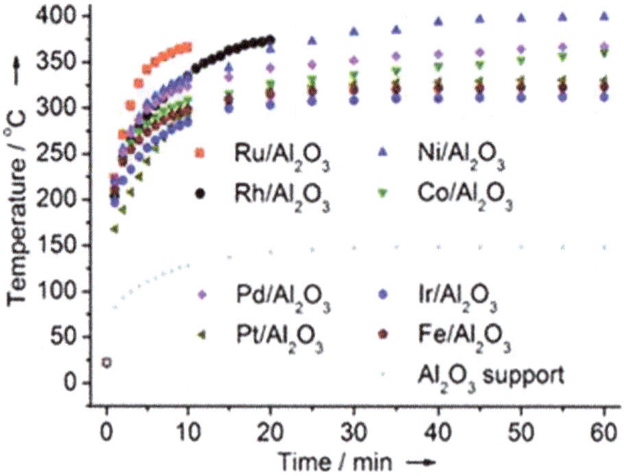

Figure 5.10: Temperature change with respect to reaction time over different catalysts [101].

A high local temperature can be achieved through the photothermal effect over the nanostructured materials having strong light harvesting ability and high specific surface area [102, 103]. This high light absorbance results from intermediate sub-bands in narrow bandgap semiconductors, including non-stoichiometric metal oxides [104, 105], metals (e.g., Au) exhibiting the localized surface plasmon resonance effect [106–109] and composite systems composed of both metallic and semiconductor materials. Cheng et al. [104] performed a defective WO_{3-x} with a mesoporous structure to photothermally reduce CO_2 into CH_4 at a rate 22-fold greater than that of WO_3 ($E_g = 2.8$ eV). A reduction in bandgap from 2.7 eV to 2.3 eV was achieved over these oxygen-vacancy-rich WO_{3-x} treated in a H_2-rich environment at various temperatures. As shown in Figure 5.11, significantly increased optical absorption was observed over these WO_{3-x} with narrow bandgaps.

Recently, a hybrid nanowires of indium oxide coated with silicon ($In_2O_{3-x}(OH_)y$/SiNW) was demonstrated to photothermally reduce CO_2 to CO at a rate of 22.0 μmol · $gcat^{-1} \cdot h^{-1}$ under simulated solar illumination (Figure 5.12) [110]. Compared with the $In_2O_{3-x}(OH)_y$ films, $In_2O_{3-x}(OH_)y$/SiNW films showed six times higher of the reaction rate. The increased catalytic performance is associated with the better utilization of both light and heat energy from solar irradiation, which facilitates the CO_2 reduction reactions. Chen et al. [111] demonstrated the light-driven catalytic CO_2 hydrogenation for producing C_{2+} hydrocarbons over an Al_2O_3-supported CoFe alloy, on which the activation of both CO_2 and H_2 occurred simultaneously. Given all the promising results

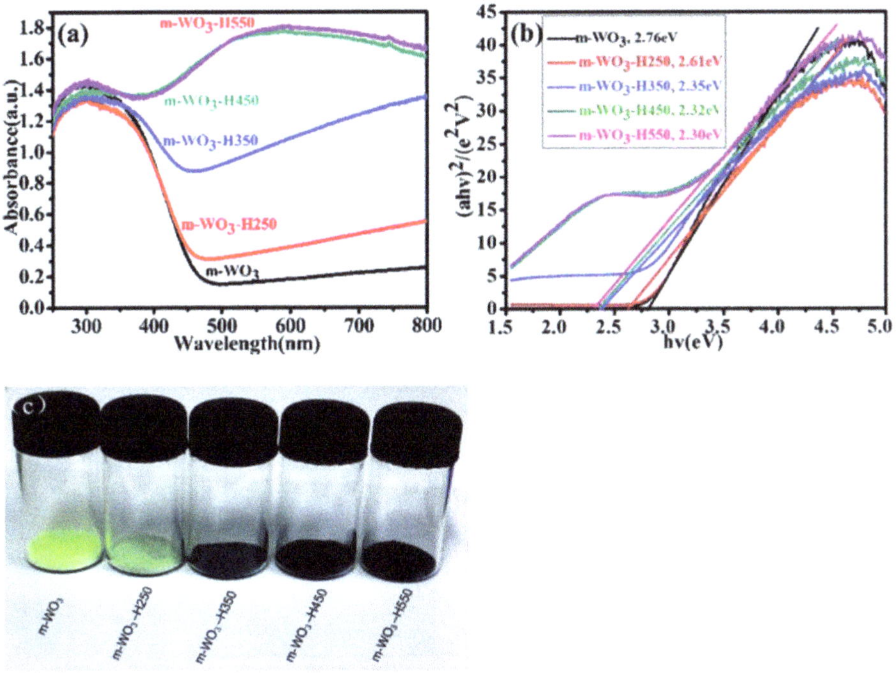

Figure 5.11: **(A)** UV/Vis spectra, **(B)** Tauc plots and **(C)** photographs of WO_3 and defective WO_{3-x} samples [104].

mentioned here, direct photothermal conversion of CO_2 provides a highly efficient and sustainable route to produce solar fuels with zero carbon dioxide emissions.

5.5 Aerosol routes to catalytic materials for CO_2 conversion

The production of nanostructured catalysts can be approached by solid, liquid, and gas methods. It is well recognized that the preparation method strongly impacts the activity of catalysts. As far, liquid-based batch processes are most frequently utilized to synthesize the catalysts owing to their simplicity. Nevertheless, they often suffer from phase separation and non-optimal catalytic performances because of the reduced active reaction surfaces from the macroscopic reactions. For the liquid-based approaches, multi-stage tedious procedures (tens of hours to days) are usually required. Furthermore, surfactant/solvent removal and calcinations are needed to further purify and crystalize the as-synthesized materials. Considering the potential scale involved in

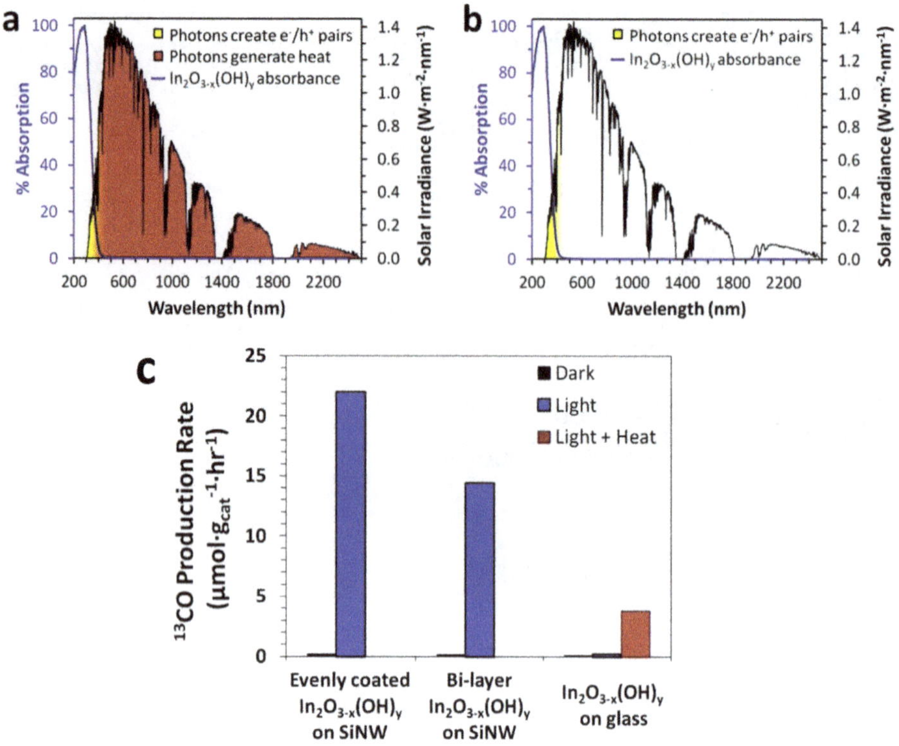

Figure 5.12: Comparison between the optical spectrum of $In_2O_{3-x}(OH)_y$ nanoparticles and the photon utilization of **(A)** $In_2O_{3-x}(OH)_y$/SiNW and **(B)** $In_2O_{3-x}(OH)_y$/glass films. **(C)** Production rates of ^{13}CO on the $In_2O_{3-x}(OH)_y$/SiNW, $In_2O_{3-x(}OH)_y$/SiNW, and $In_2O_{3-x}(OH)_y$/glass under various conditions [110].

producing nanostructured catalysts for CO_2 conversion, a high-throughput and fast process for making high-performance catalysts is highly desirable.

Recently gas-phase (aerosol) synthetic routes have emerged as a powerful strategy for material engineering due to its fast and simple synthesis with high throughput as compared to the traditional liquid or solid methods. The gas-phase routes include flame spray synthesis, plasma methods, chemical vapor deposition, and spray pyrolysis/drying, etc. The processing of materials via the aerosol processes has been extensively researched over a wide range of applications, including catalysts, adsorbents, optical fibers, sensors, etc. But so far, the research works related to the field of CO_2 conversion are still very limited. In the following sections, special emphases are on the aerosol-synthesized catalysts being studied in CO_2 conversion.

5.5.1 Aerosol chemical vapor deposition (ACVD)

Aerosol chemical vapor deposition is a gas-to-solid conversion process, where materials are generated by cooling a supersaturated vapor via chemical vapor deposition [112]. Wang and co-workers [113] synthesized the Pt-TiO_2 films using an ACVD route to deposit the columnar TiO_2 film, followed by a tilted-target sputtering technique to decorate Pt NPs on the preformed TiO_2. The crystalline structure (anatase and rutile), and morphology (dense, columnar and granular) of the TiO_2 are modulated by changing the reaction temperature, reaction time and residence time of the system, as displayed inFigure 5.13(a). Particularly, the columnar TiO_2 could prolong the lifetime of the photo-induced charge carriers, leading to a superior photocatalytic performance to other morphologies. The quantum yield of the aerosol-made thin film exhibited a value of 2.41%, yields higher than the reported values in the literature of CO_2 photoreduction. Meanwhile, the change in energy levels of the Pt NPs in the light of their size also has a strong impact on its CO_2 photoreduction performance (Figure 5.13(b)). When the size of Pt is larger than 1 nm, it simply acts as a recombination site which traps the charge carriers and results in a decreased photocatalytic efficiency. On the contrary, when the size of Pt is smaller than 1 nm, quantum confinement effects make its energy band separation too high for an electron to be travelled to the Pt.

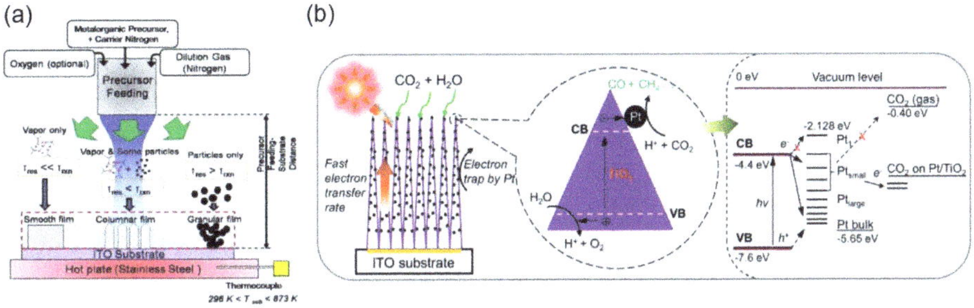

Figure 5.13: **(A)** ACVD for TiO_2 formation with various morphologies; **(B)** Schematics of CO_2 reduction on the Pt-TiO_2 films [113].

5.5.2 Aerosol spray pyrolysis and self-assembly

Aerosol spray pyrolysis (ASP) is a liquid-to-solid conversion process, in which the atomized precursor droplets serve as microreactors for particle formation [114]. In this system, droplet formation, precipitation, decomposition/oxidation, and sintering stages can be integrated into a single continuous process. The ASP has been extensively applied in preparing various mixed oxides and metal oxide-supported metal catalysts, such as Pt/Al_2O_3 and Pt/$CeZrO_2$. Liu and co-workers showed that

ASP is effective in producing MgO-modified TiO_2 composites (Figure 5 14) that photocatalytically reduce the CO_2 [21]. They showed that highly aggregated MgO particles were formed on the surface of the ASP-synthesized Mg/Ti-SP, while uniformly dispersed MgO were observed on the Mg/Ti-WI prepared by conventional wet impregnation (WI) method. Interestingly, more than two times higher CO production yield was observed over the Mg/Ti-SP than that of the Mg/Ti-WI. Furthermore, the Mg/Ti-SP catalyst was more deactivation-resistant compared to the Mg/Ti-WI. The *in-situ* surface intermediate analyses on the adsorption and photoreduction of CO_2 revealed that the surface morphology of MgO/TiO_2 composites had strong impacts on the separation efficiency of charge carrier and formation of different surface intermediates; both of which together led to a higher activity and stability of the Mg/Ti-SP, providing a commercially viable route to fabricate effective and long-lasting catalyst for solar fuels production.

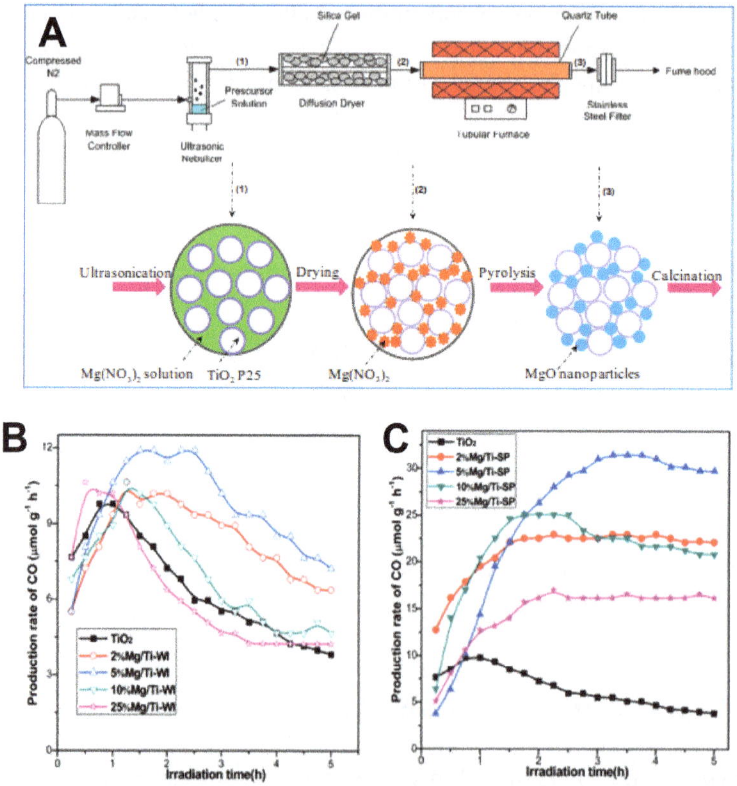

Figure 5.14: **(A)** Formation of MgO/TiO_2 via the SP process. The CO production rate on **(B)** Mg/Ti-WI and **(C)** Mg/Ti-SP with a different Mg concentration under UV-vis light irradiation at 150 °C [110].

In addition to the preparation of metal oxide-based catalysts, non-metal catalytic materials can also be achieved in this gas-phase system. For example, Lin et al. [23] recently demonstrated the fabrication of carbon dots in an aerosol reactor. Carbon dots are promising candidates for a wide range of photocatalytic applications due to their unique photoluminescent and up-conversion properties [115, 116]. Generally, carbon dots are prepared via solution-based routes, in which multi-steps including hydrolysis and thermolysis of carbon precursors, collection and purification and as-prepared carbon dots are involved. In this aerosol system, the carbon precursor (e.g., citric acid) was rapidly pyrolyzed in the aerosol microdroplets and small and uniform carbon dots (~2 nm) were formed due to the fast drying/pyrolysis induced by the droplet's high surface-to-volume ratio. Further, a hollow composite consisting nonstoichiometric ZnO_{1-x}/carbon dots was also achieved via FuAR (Figure 5.15. **(A-E)**), and showed a high CO production rate under near-infrared light only (Figure 5.15. **(F-G)**).

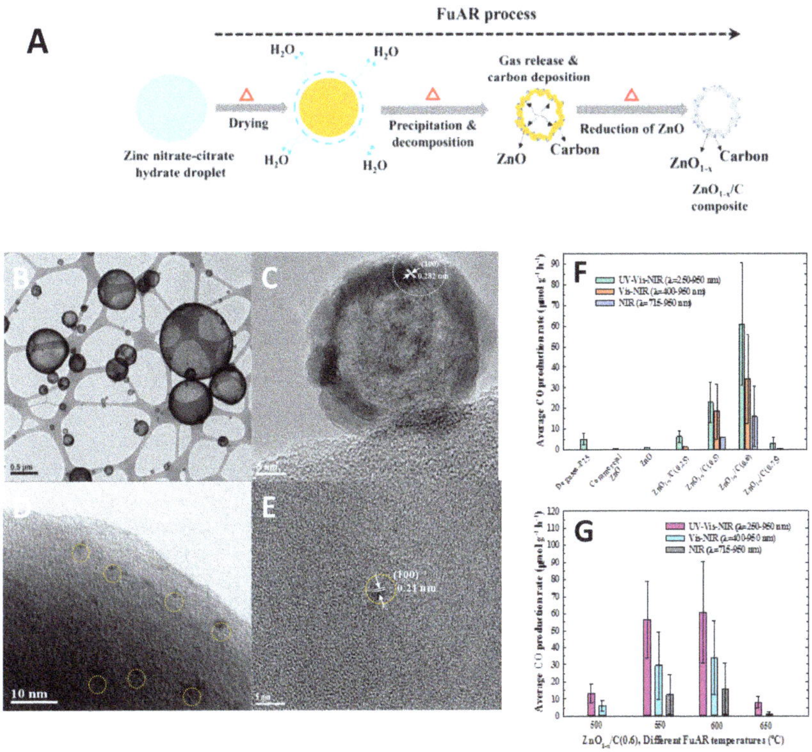

Figure 5.15: **(A)** Scheme of the formation of the ZnO_{1-x}/carbon dots composite HSs. **(B-C)** TEM images of the ZnO_{1-x}/carbon dots HSs, **(D-E)** HRTEM images of the Zn-free carbon dots and **(F-G)** The influences of precursor molar ratio and reaction temperature on the CO production from CO_2 photoreduction on ZnO_{1-x}/carbon dots HSs [23].

Besides aerosol spray pyrolysis, aerosol-assisted self-assembly (AASA) is another gas-phase technique used to prepare the particles with hierarchical structures at the micro- and nanoscales, such as porous, hollow, and other nanostructures [117]. Wang and co-workers used a spray approach to produce $Cu/TiO_2/SiO_2$ mesoporous composites for photoreducing CO_2 into CO [118]. As illustrated in Figure 5.16, in this method, Cu nitrate, SiO_2, and TiO_2 colloids in water were well mixed to create a homogeneous suspension, which was then atomized, delivered into a pre-heated furnace. Because of the rapid heating during the aerosol process, these composite particles were formed within several seconds; more importantly, the physicochemical properties (e.g., composition and porosity) of the particles can be finely controlled by adjusting the synthetic factors. The CO_2 photoreduction tests showed that the $Cu/TiO_2/SiO_2$ particles had the highest CO yield of approximately 20 mmol g_{TiO2}^{-1} h^{-1}, when the percentages of TiO_2 and Cu equal to 2 mol% and 0.01 mol%, respectively.

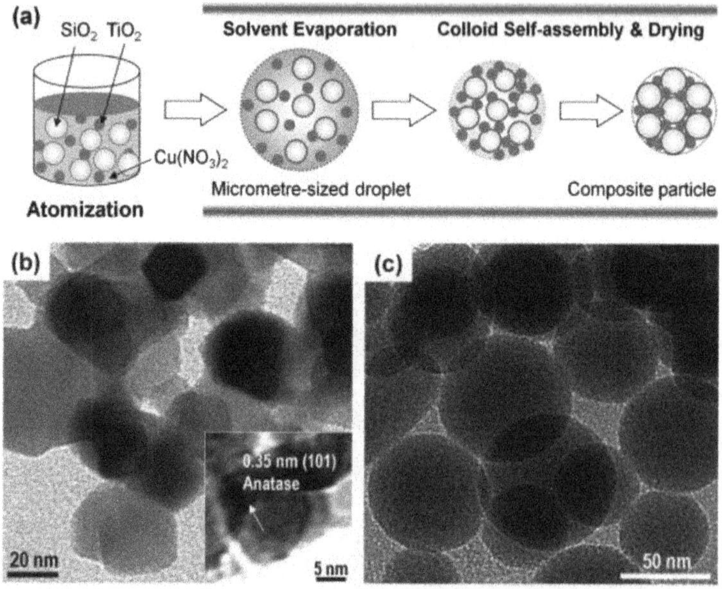

Figure 5.16: Production of $Cu/TiO_2/SiO_2$ particles via AASA. **(A)** Formation mechanism and TEM images of **(B)** TiO_2 and **(C)** SiO_2 nanocolloids [118].

5.5.3 Flame spray pyrolysis

Flame spray pyrolysis (FSP) is a continuous aerosol process that generates nanoparticles in one step with high throughput and good control of different structural parameters [119–121]. In this system, the metal precursors in the gas or liquid phase are fed through a nozzle to produce aerosols, which are delivered to a high-temperature or plasma environment (Figure 5.17). The heated precursors undergo evaporation, nucleation, and coagulation to form the final particles. A variety of nanostructured materials produced via the FSP processes have been extensively reviewed within the past decades [122–124]. Tada and co-workers [125] used a FSP approach for preparing a series of CuO-ZrO_2 catalysts for hydrogenation of CO_2. A finely tailored size of CuO was achieved by changing the feeding rate of oxygen flow, and methanol production was favored over the CuO–ZrO_2 catalysts with smaller CuO sizes. More recently, Schubert and co-workers [126] explored the use a two-nozzle FSP for preparing the CoO_x–Al_2O_3 catalysts, and compared with their counterparts for CO_2 methanation. In this FSP system, single component of Co_3O_4 and Al_2O_3 or $CoAl_2O_4$ composites could be realized by manipulating the angle between the two nozzles, revealing its great potential in precisely modulating the composition of catalysts.

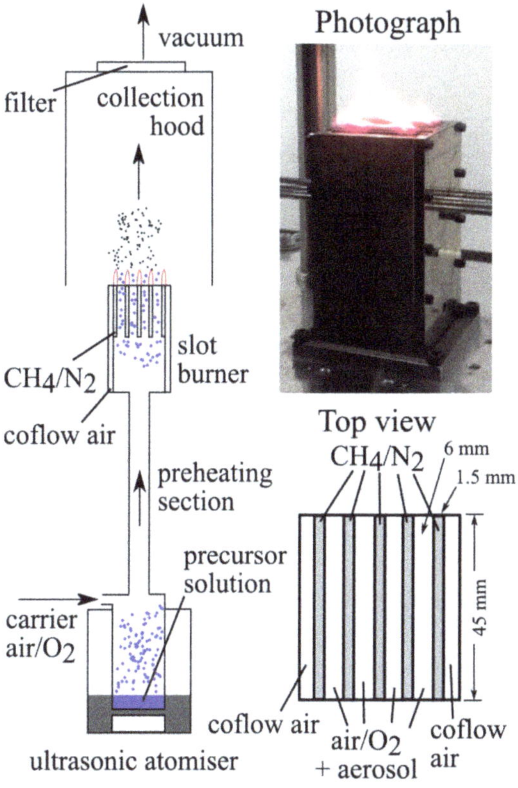

Figure 5.17: The schematic of the aerosol flame system [121].

5.6 Conclusions and outlook

Preceding energy-intensive catalytic processes using utilizing solar energy are beneficial in terms of environmental and economic impacts. In this chapter, recent works on photo(electro)chemical and photothermal processes for CO_2 conversion are summarized. The integration of photochemistry with electrochemistry or thermochemistry offers great potentials in achieving high energy efficiency and high rates of CO_2 conversion. While the key research objective is to design a catalytic material that simultaneously drives the photo(electro)chemical and photothermal reactions for CO_2 conversion with high activity and selectivity, future works should be directed to explore the reduction mechanisms of photo(electro)catalytic and photothermal CO_2 conversion, and to understand their synergistic effects on CO_2 conversion. The reaction device configuration to maximize the utilization of photo(electro) chemical and photothermal energies must be developed as well.

A rapid growth in the number of works has recently been conducted on applying aerosol-made nanocatalysts in conventional heterogeneous catalysis such as oxidation of CO and hydrocarbons, reduction of nitrogen oxides and degradation of aquatic contaminants. Nevertheless, the discovery and fine control of properties of novel aerosol-made catalysts for CO_2 conversion, and their corresponding mechanistic studies are still very rare. In comparison to the catalytic materials produced via the solution-based processes, aerosol-based routes are expected to be a potentially economic method for producing catalysts at an industrial scale. This has been particularly important for environmental pollutant elimination in realistic environments in which a great amount of catalysts are needed.

So far, the efficiencies of solar fuel generation by light-driven heterogeneous catalysis are far from satisfaction, and much work remains to be addressed before this method can be considered economically feasible. With in-depth understanding of fundamental structure-composition-activity relationships, the physicochemical properties of catalysts are expected to be precisely modulated to optimize the catalytic performances. In this regard, computational studies and modern *in-situ* characterization techniques are thus highly required to provide more in-depth information on physicochemical properties for better design of functional catalysts.

References

[1] Koytsoumpa, E.I., Bergins, C., Kakaras, E. (2018). The CO_2 economy: Review of CO_2 capture and reuse technologies. *Journal of Supercritical Fluids* 132: 3–16.

[2] Zheng, Y., Zhang, W., Li, Y., Chen, J., Yu, B., Wang, J., Zhang, L., Zhang, J. (2017). Energy related CO_2 conversion and utilization: Advanced materials/nanomaterials, reaction mechanisms and technologies. *Nano Energy* 40: 512–539.

[3] Luis, P. (2016). Use of monoethanolamine (MEA) for CO_2 capture in a global scenario: Consequences and alternatives. *Desalination* 380: 93–99.
[4] Wang, M., Joel, A.S., Ramshaw, C., Eimer, D., Musa, N.M. (2015). Process intensification for post-combustion CO_2 capture with chemical absorption: A critical review. *Applied Energy* 158: 275–291.
[5] Oh, S.-Y., Binns, M., Cho, H., Kim, J.-K. (2016). Energy minimization of MEA-based CO_2 capture process. *Applied Energy* 169: 353–362.
[6] Bachu, S. (2015). Review of CO_2 storage efficiency in deep saline aquifers. *International Journal of Greenhouse Gas Control* 40: 188–202.
[7] Jarvis, S.M., Samsatli, S. (2018). Technologies and infrastructures underpinning future CO_2 value chains: A comprehensive review and comparative analysis. *Renewable & Sustainable Energy Reviews* 85: 46–68.
[8] Stuardi, F.M., MacPherson, F., Leclaire, J. (2019). Integrated CO_2-capture and utilization: A priority research direction. *Current Opinion in Green and Sustainable Chemistry*.
[9] Cui, Y., Lian, X., Xu, L., Chen, M., Yang, B., Wu, C., Li, W., Huang, B., Hu, X. (2019). Designing and fabricating ordered mesoporous metal oxides for CO_2 catalytic conversion: A review and prospect. *Materials* 12: 276.
[10] Xie, S., Zhang, Q., Liu, G., Wang, Y. (2016). Photocatalytic and photoelectrocatalytic reduction of CO_2 using heterogeneous catalysts with controlled nanostructures. *Chemical Communications* 52: 35–59.
[11] Castro, S., Albo, J., Irabien, A. (2018). Photoelectrochemical Reactors for CO_2 Utilization. *ACS Sustainable Chemistry & Engineering* 6: 15877–15894.
[12] Wang, W.-N., Soulis, J., Yang, Y.J., Biswas, P. (2014). Comparison of CO_2 photoreduction systems: A review. *Aerosol and Air Quality Research* 14: 533–549.
[13] Kho, E.T., Tan, T.H., Lovell, E., Wong, R.J., Scott, J., Amal, R. (2017). A review on photo-thermal catalytic conversion of carbon dioxide. *Green Energy Environment* 2: 204–217.
[14] Tang, S., Sun, J., Hong, H., Liu, Q. (2017). Solar fuel from photo-thermal catalytic reactions with spectrum-selectivity: A review. *Frontiers Energy* 11: 437–451.
[15] Xu, C., Zhang, Y., Pan, F., Huang, W., Deng, B., Liu, J., Wang, Z., Ni, M., Cen, K. (2017). Guiding effective nanostructure design for photo-thermochemical CO_2 conversion: From DFT calculations to experimental verifications. *Nano Energy* 41: 308–319.
[16] Ola, O., Maroto-Valer, M.M. (2015). Review of material design and reactor engineering on TiO_2 photocatalysis for CO_2 reduction. *Journal of Photochemistry and Photobiology C: Photochemistry Reviews* 24: 16–42.
[17] Chien Nguyen, C., Nang Vu, N., Do, T.-O. (2015). Recent advances in the development of sunlight-driven hollow structure photocatalysts and their applications. *Journal of Materials Chemistry A* 3: 18345–18359.
[18] Li, K., An, X., Park, K.H., Khraisheh, M., Tang, J. (2014). A critical review of CO_2 photoconversion: Catalysts and reactors. *Catalysis Today* 224: 3–12.
[19] Yu, J., Jin, J., Cheng, B., Jaroniec, M. (2014). A noble metal-free reduced graphene oxide–CdS nanorod composite for the enhanced visible-light photocatalytic reduction of CO_2 to solar fuel. *Journal of Materials Chemistry A* 2: 3407–3416.
[20] Zhao, H., Liu, L., Andino, J.M., Li, Y. (2013). Bicrystalline TiO_2 with controllable anatase–brookite phase content for enhanced CO_2 photoreduction to fuels. *Journal of Materials Chemistry A* 1: 8209–8216.
[21] Liu, L., Zhao, C., Pitts, D., Zhao, H., Li, Y. (2014). CO_2 photoreduction with H_2O vapor by porous MgO–TiO_2 microspheres: Effects of surface MgO dispersion and CO_2 adsorption–desorption dynamics. *Catalysis Science and Technology* 4: 1539–1546.

[22] Tahir, M., Amin, N.S. (2016). Performance analysis of nanostructured $NiO-In_2O_3/TiO_2$ catalyst for CO_2 photoreduction with H_2 in a monolith photoreactor. *Chemical Engineering Journal* 285: 635–649.
[23] Lin, L.-Y., Kavadiya, S., Karakocak, B.B., Nie, Y., Raliya, R., Wang, S.T., Berezin, M.Y., Biswas, P. (2018). ZnO_{1-x}/carbon dots composite hollow spheres: Facile aerosol synthesis and superior CO_2 photoreduction under UV, visible and near-infrared irradiation. *Applied Catalysis B: Environmental* 230: 36–48.
[24] Linsebigler, A.L., Lu, G., Yates, J.T. (1995). Photocatalysis on TiO_2 surfaces: Principles, mechanisms, and selected results. *Chemical Reviews* 95: 735–758.
[25] Inoue, T., Fujishima, A., Konishi, S., Honda, K. (1979). Photoelectrocatalytic reduction of carbon dioxide in aqueous suspensions of semiconductor powders. *Nature* 277: 637.
[26] Kubacka, A., Fernández-García, M., Colón, G. (2012). Advanced nanoarchitectures for solar photocatalytic applications. *Chemical Reviews* 112: 1555–1614.
[27] Pan, Y.-X., You, Y., Xin, S., Li, Y., Fu, G., Cui, Z., Men, Y.-L., Cao, -F.-F., Yu, S.-H., Goodenough, J.B. (2017). Photocatalytic CO_2 reduction by carbon-coated indium-oxide nanobelts. *Journal of the American Chemical Society* 139: 4123–4129.
[28] Huang, Z., Teramura, K., Asakura, H., Hosokawa, S., Tanaka, T. (2017). Efficient photocatalytic carbon monoxide production from ammonia and carbon dioxide by the aid of artificial photosynthesis. *Chemical Science* 8: 5797–5801.
[29] Huang, Z., Yoshizawa, S., Teramura, K., Asakura, H., Hosokawa, S., Tanaka, T. (2018). Photocatalytic conversion of carbon dioxide over A2BTa5O15 (A = Sr, Ba; B = K, Na) using ammonia as an efficient sacrificial reagent. *ACS Sustainable Chemistry & Engineering* 6: 8247–8255.
[30] Nahar, S., Zain, M.F.M., Kadhum, A.A.H., Hasan, H.A., Hasan, M.R. (2017). Advances in photocatalytic CO_2 reduction with water: A review. *Materials* 10: 629.
[31] Yang, Z., Wei, J., Zeng, G., Zhang, H., Tan, X., Ma, C., Li, X., Li, Z., Zhang, C. (2019). A review on strategies to LDH-based materials to improve adsorption capacity and photoreduction efficiency for CO_2, *Coord. Chemical Reviews* 386: 154–182.
[32] Low, J., Cheng, B., Yu, J. (2017). Surface modification and enhanced photocatalytic CO_2 reduction performance of TiO_2: A review. *Applied Surface Science* 392: 658–686.
[33] Wang, Y., Zhao, J., Wang, T., Li, Y., Li, X., Yin, J., Wang, C. (2016). CO_2 photoreduction with H_2O vapor on highly dispersed CeO_2/TiO_2 catalysts: Surface species and their reactivity. *Journal of Catalysis* 337: 293–302.
[34] Li, X., Liu, H., Luo, D., Li, J., Huang, Y., Li, H., Fang, Y., Xu, Y., Zhu, L. (2012). Adsorption of CO_2 on heterostructure $CdS(Bi_2S_3)/TiO_2$ nanotube photocatalysts and their photocatalytic activities in the reduction of CO_2 to methanol under visible light irradiation. *Chemical Engineering Journal* 180: 151–158.
[35] Huang, Q., Yu, J., Cao, S., Cui, C., Cheng, B. (2015). Efficient photocatalytic reduction of CO_2 by amine-functionalized g-C_3N_4. *Applied Surface Science* 350–355.
[36] Fang, B., Xing, Y., Bonakdarpour, A., Zhang, S., Wilkinson, D.P. (2015). Hierarchical $CuO-TiO_2$ hollow microspheres for highly efficient photodriven reduction of CO_2 to CH_4. ACS *Sustainable Chemistry & Engineering* 3: 2381–2388.
[37] Zhang, Q., Gao, T., Andino, J.M., Li, Y. (2012). Copper and iodine co-modified TiO_2 nanoparticles for improved activity of CO_2 photoreduction with water vapor. *Applied Catalysis B: Environmental* 123–124: 257–264.
[38] Li, X., Zhuang, Z., Li, W., Pan, H. (2012). Photocatalytic reduction of CO_2 over noble metal-loaded and nitrogen-doped mesoporous TiO_2. *Applied Catalysis A: General* 429–430: 31–38.

[39] Pham, T.-D., Lee, B.-K. (2017). Novel photocatalytic activity of Cu@V co-doped TiO_2/PU for CO_2 reduction with H_2O vapor to produce solar fuels under visible light. *Journal of Catalysis* 345: 87–95.
[40] Wang, T., Meng, X., Liu, G., Chang, K., Li, P., Kang, Q., Liu, L., Li, M., Ouyang, S., Ye, J. (2015). In situ synthesis of ordered mesoporous Co-doped TiO_2 and its enhanced photocatalytic activity and selectivity for the reduction of CO_2. *Journal of Materials Chemistry A* 3: 9491–9501.
[41] Liu, L., Jiang, Y., Zhao, H., Chen, J., Cheng, J., Yang, K., Li, Y. (2016). Engineering coexposed {001} and {101} facets in oxygen-deficient TiO_2 nanocrystals for enhanced CO_2 photoreduction under visible light. *ACS Catalysis* 6: 1097–1108.
[42] Tan, -L.-L., Ong, W.-J., Chai, S.-P., Mohamed, A.R. (2016). Visible-light-activated oxygen-rich TiO_2 as next generation photocatalyst: Importance of annealing temperature on the photoactivity toward reduction of carbon dioxide. *Chemical Engineering Journal* 283: 1254–1263.
[43] Xie, K., Umezawa, N., Zhang, N., Reunchan, P., Zhang, Y., Ye, J. (2011). Self-doped $SrTiO_{3-\delta}$ photocatalyst with enhanced activity for artificial photosynthesis under visible light. *Energy and Environmental Sciences* 4: 4211–4219.
[44] Li, K., Peng, T., Ying, Z., Song, S., Zhang, J. (2016). Ag-loading on brookite TiO_2 quasi nanocubes with exposed {2 1 0} and {0 0 1} facets: Activity and selectivity of CO_2 photoreduction to CO/CH_4. *Applied Catalysis B: Environmental* 180: 130–138.
[45] Liu, L., Gao, F., Zhao, H., Li, Y. (2013). Tailoring Cu valence and oxygen vacancy in Cu/TiO_2 catalysts for enhanced CO_2 photoreduction efficiency. *Applied Catalysis B: Environmental* 134–135: 349–358.
[46] Zhao, C., Krall, A., Zhao, H., Zhang, Q., Li, Y. (2012). Ultrasonic spray pyrolysis synthesis of Ag/TiO_2 nanocomposite photocatalysts for simultaneous H2 production and CO_2 reduction. *International Journal of Hydrogen Energy* 37: 9967–9976.
[47] Wang, Y., Zhao, J., Li, Y., Wang, C. (2018). Selective photocatalytic CO_2 reduction to CH_4 over Pt/$In_2$03: Significant role of hydrogen adatom. *Applied Catalysis B: Environmental* 226: 544–553.
[48] Umezawa, N., Kristoffersen, H.H., Vilhelmsen, L.B., Hammer, B. (2016). Reduction of CO_2 with water on Pt-Loaded Rutile TiO_2(110) modeled with density functional theory. *Journal of Physical Chemistry C* 120: 9160–9164.
[49] Zhao, H., Pan, F., Li, Y. (2017). A review on the effects of TiO_2 surface point defects on CO_2 photoreduction with H_2O. *Journal of Materiomics* 3: 17–32.
[50] Xie, S., Wang, Y., Zhang, Q., Fan, W., Deng, W., Wang, Y. (2013). Photocatalytic reduction of CO_2 with H_2O : Significant enhancement of the activity of Pt–TiO_2 in CH_4 formation by addition of MgO. *Chemical Communications* 49: 2451–2453.
[51] Tahir, M., Tahir, B., Amin, N.A.S. (2015). Gold-nanoparticle-modified TiO_2 nanowires for plasmon-enhanced photocatalytic CO_2 reduction with H_2 under visible light irradiation. *Applied Surface Science* 356: 1289–1299.
[52] Tahir, M., Tahir, B., Amin, N.A.S. (2017). Synergistic effect in plasmonic Au/Ag alloy NPs co-coated TiO_2 NWs toward visible-light enhanced CO_2 photoreduction to fuels. *Applied Catalysis B: Environmental* 204: 548–560.
[53] Low, J., Qiu, S., Xu, D., Jiang, C., Cheng, B. (2018). Direct evidence and enhancement of surface plasmon resonance effect on Ag-loaded TiO_2 nanotube arrays for photocatalytic CO_2 reduction. *Applied Surface Science* 434: 423–432.
[54] Cheng, X., Dong, P., Huang, Z., Zhang, Y., Chen, Y., Nie, X., Zhang, X. (2017). Green synthesis of plasmonic Ag nanoparticles anchored TiO_2 nanorod arrays using cold plasma for visible-light-driven photocatalytic reduction of CO_2. *Journal of CO2 Utilization* 20: 200–207.

[55] Xiang, Q., Yu, J., Jaroniec, M. (2012). Graphene-based semiconductor photocatalysts. *Chemical Society Reviews* 41: 782–796.
[56] Zhu, J., Chen, M., He, Q., Shao, L., Wei, S., Guo, Z. (2013). An overview of the engineered graphene nanostructures and nanocomposites. *RSC Advances* 3: 22790–22824.
[57] Xiang, Q., Cheng, B., Yu, J. (2015). Graphene-based photocatalysts for solar-fuel generation. *Angewandte Chemie International Edition* 54: 11350–11366.
[58] Sadeghi, N., Sharifniaa, S., Do, T.-O. (2018). Enhanced CO_2 photoreduction by a graphene–porphyrin metal–organic framework under visible light irradiation. *Journal of Materials Chemistry A* 6: 18031–18035.
[59] Low, J., Yu, J., Ho, W. (2015). Graphene-based photocatalysts for CO_2 reduction to solar fuel. The *Journal of Physical Chemistry Letters* 6: 4244–4251.
[60] Kim, H., Moon, G., Monllor-Satoca, D., Park, Y., Choi, W. (2012). Solar photoconversion using graphene/TiO_2 composites: Nanographene shell on TiO_2 core versus TiO_2 nanoparticles on graphene sheet. *Journal of Physical Chemistry C* 116: 1535–1543.
[61] Zhang, Y., Tang, Z.-R., Fu, X., Xu, Y.-J. (2010). TiO_2–graphene nanocomposites for gas-phase photocatalytic degradation of volatile aromatic pollutant: Is TiO_2–graphene truly different from other TiO_2–carbon composite materials? *ACS Nano* 4: 7303–7314.
[62] Hou, J., Cheng, H., Takeda, O., Zhu, H. (2015). Three-dimensional bimetal-graphene-semiconductor coaxial nanowire arrays to harness charge flow for the photochemical reduction of carbon dioxide. *Angewandte Chemie International Edition* 54: 8480–8484.
[63] Lin, L.-Y., Nie, Y., Kavadiya, S., Soundappan, T., Biswas, P. (2017). N-doped reduced graphene oxide promoted nano TiO_2 as a bifunctional adsorbent/photocatalyst for CO_2 photoreduction: Effect of N species. *Chemical Engineering Journal* 316: 449–460.
[64] Sun, S., Watanabe, M., Wang, P., Ishihara, T. (2019). Synergistic enhancement of H_2 and CH_4 evolution by CO_2 photoreduction in water with reduced graphene oxide–bismuth monoxide quantum dot catalyst. *ACS Applied Energy Materials* 2: 2104–2112.
[65] Li, H., Song, X., Shi, Y., Gao, Y., Si, D., Hao, C. (2019). Role of water oxidation in the photoreduction of graphene oxide. *Chemical Communications* 55: 1837–1840.
[66] Liang, Y.T., Vijayan, B.K., Gray, K.A., Hersam, M.C. (2011). Minimizing graphene defects enhances titania nanocomposite-based photocatalytic reduction of CO_2 for improved solar fuel production. *Nano Letters* 11: 2865–2870.
[67] Jung, H., Cho, K.M., Kim, K.H., Yoo, H.-W., Al-Saggaf, A., Gereige, I. and Jung, H.-T. (2018). Highly efficient and stable CO_2 reduction photocatalyst with a hierarchical structure of mesoporous TiO_2 on 3D graphene with few-layered MoS_2. *ACS Sustainable Chemistry & Engineering* 6: 5718–5724.
[68] Nie, Y., Wang, W.-N., Jiang, Y., Fortner, J. and Biswas, P. (2016). Crumpled reduced graphene oxide–amine–titanium dioxide nanocomposites for simultaneous carbon dioxide adsorption and photoreduction. *Catalysis Science and Technology* 6187–6196.
[69] Cho, K.M., Kim, K.H., Park, K., Kim, C., Kim, S., Al-Saggaf, A., Gereige, I., Jung, H.-T. (2017). Amine-functionalized graphene/CdS composite for photocatalytic reduction of CO_2. *ACS Catalysis* 7: 7064–7069.
[70] Hsu, H.-C., Shown, I., Wei, H.-Y., Chang, Y.-C., Du, H.-Y., Lin, Y.-G., Tseng, C.-A., Wang, C.-H., Chen, L.-C., Lin, Y.-C., Chen, K.-H. (2013). Graphene oxide as a promising photocatalyst for CO_2 to methanol conversion. *Nanoscale* 5: 262–268.
[71] Wang, -C.-C., Zhang, Y.-Q., Li, J., Wang, P. (2015). Photocatalytic CO_2 reduction in metal–organic frameworks: A mini review. *Journal of Molecular Structure* 1083: 127–136.
[72] Wang, X.-K., Liu, J., Zhang, L., Dong, L.-Z., Li, S.-L., Kan, Y.-H., Li, D.-S., Lan, Y.-Q. (2019). Monometallic catalytic models hosted in stable metal–organic frameworks for tunable CO_2 photoreduction. *ACS Catalysis* 9: 1726–1732.

[73] Chen, D., Xing, H., Wang, C., Su, Z. (2016). Highly efficient visible-light-driven CO_2 reduction to formate by a new anthracene-based zirconium MOF via dual catalytic routes. *Journal of Materials Chemistry A* 4: 2657–2662.
[74] Zhang, H., Wei, J., Dong, J., Liu, G., Shi, L., An, P., Zhao, G., Kong, J., Wang, X., Meng, X., Zhang, J., Ye, J. (2016). Efficient visible-light-driven carbon dioxide reduction by a single-atom implanted metal–organic framework. *Angewandte Chemie International Edition* 55: 14310–14314.
[75] Li, R., Zhang, W., Zhou, K. (2018). Metal–organic-framework-based catalysts for photoreduction of CO_2. *Advanced Materials* 30: 1705512.
[76] Han, B., Ou, X., Deng, Z., Song, Y., Tian, C., Deng, H., Xu, Y.-J. and Lin, Z. (2018). Nickel metal–organic framework monolayers for photoreduction of diluted CO_2: Metal-node-dependent activity and selectivity. *Angewandte Chemie International Edition* 57: 16811–16815.
[77] Xu, H.-Q., Hu, J., Wang, D., Li, Z., Zhang, Q., Luo, Y., Yu, S.-H., Jiang, H.-L. (2015). Visible-light photoreduction of CO_2 in a metal–organic framework: Boosting electron–hole separation via electron trap states. *Journal of the American Chemical Society* 137: 13440–13443.
[78] Crake, A., Christoforidis, K.C., Gregg, A., Moss, B., Kafizas, A., Petit, C. (2019). The effect of materials architecture in TiO_2/MOF composites on CO_2 photoreduction and charge transfer. *Small* 15: 1805473.
[79] He, X., Gan, Z., Fisenko, S., Wang, D., El-Kaderi, H.M., Wang, W.-N. (2017). Rapid formation of metal-organic frameworks (MOFs)-based nanocomposites in microdroplets and their applications for CO_2 photoreduction. *ACS Applied Materials & Interfaces* 9: 9688–9698.
[80] Li, R., Hu, J., Deng, M., Wang, H., Wang, X., Hu, Y., Jiang, H.-L., Jiang, J., Zhang, Q., Xie, Y., Xiong, Y. (2014). Integration of an inorganic semiconductor with a metal–organic framework: A platform for enhanced gaseous photocatalytic reactions. *Advanced Materials* 26: 4783–4788.
[81] Tahir, M., Amin, N.S. (2013). Photocatalytic CO_2 reduction and kinetic study over In/TiO_2 nanoparticles supported microchannel monolith photoreactor. *Applied Catalysis A: General* 467: 483–496.
[82] Tan, S.S., Zou, L., Hu, E. (2008). Kinetic modelling for photosynthesis of hydrogen and methane through catalytic reduction of carbon dioxide with water vapour. *Catalysis Today* 131: 125–129.
[83] Tan, -L.-L., Ong, W.-J., Chai, S.-P., Mohamed, A.R. (2017). Photocatalytic reduction of CO_2 with H_2O over graphene oxide-supported oxygen-rich TiO_2 hybrid photocatalyst under visible light irradiation: Process and kinetic studies. *Chemical Engineering Journal* 308: 248–255.
[84] Castro, S., Albo, J., Irabien, A. (2018). Photoelectrochemical reactors for CO_2 utilization. ACS Sustainable *Chemistry & Engineering* 6: 15877–15894.
[85] Sato, S., Arai, T., Morikawa, T., Uemura, K., Suzuki, T.M., Tanaka, H., Kajino, T. (2011). Selective CO_2 conversion to formate conjugated with H_2O oxidation utilizing semiconductor/complex hybrid photocatalysts. *Journal of the American Chemical Society* 133: 15240–15243.
[86] Yu, Y., Zhang, Z., Yin, X., Kvit, A., Liao, Q., Kang, Z., Yan, X., Zhang, Y., Wang, X. (2017). Enhanced photoelectrochemical efficiency and stability using a conformal TiO_2 film on a black silicon photoanode. *Nature of Energy* 2: 17045.
[87] Zhao, J., Wang, X., Xu, Z., Loo, J.S.C. (2014). Hybrid catalysts for photoelectrochemical reduction of carbon dioxide: A prospective review on semiconductor/metal complex co-catalyst systems. *Journal of Materials Chemistry A* 2: 15228–15233.
[88] Sahara, G., Kumagai, H., Maeda, K., Kaeffer, N., Artero, V., Higashi, M., Abe, R., Ishitani, O. (2016). Photoelectrochemical reduction of CO_2 coupled to water oxidation using a

photocathode with a Ru(II)–Re(I) complex photocatalyst and a CoO_x/TaON photoanode. *Journal of the American Chemical Society* 138: 14152–14158.
[89] Chen, W., Han, B., Tian, C., Liu, X., Liang, S., Deng, H., Lin, Z. (2019). MOFs-derived ultrathin holey Co_3O_4 nanosheets for enhanced visible light CO_2 reduction. *Applied Catalysis B: Environmental* 244: 996–1003.
[90] Heidary, N., Harris, T.G.A.A., Ly, K.H., Kornienko, N. Artificial photosynthesis with metal and covalent organic frameworks (MOFs and COFs): Challenges and prospects in fuel-forming electrocatalysis. *Physiologia Plantarum*.
[91] Liu, -S.-S., Xing, Q.-J., Chen, Y., Zhu, M., Jiang, X.-H., Wu, S.-H., Dai, W., Zou, J.-P. (2019). Photoelectrochemical degradation of organic pollutants using BiOBr anode coupled with simultaneous CO_2 reduction to liquid fuels via CuO cathode. *ACS Sustainable Chemistry & Engineering* 7: 1250–1259.
[92] Cheng, J., Xuan, X., Yang, X., Zhou, J., Cen, K. (2019). Selective reduction of CO_2 to alcohol products on octahedral catalyst of carbonized Cu(BTC) doped with Pd nanoparticles in a photoelectrochemical cell. *Chemical Engineering Journal* 358: 860–868.
[93] Zhang, W., Li, R., Zhao, X., Chen, Z., Law, A.W.-K., Zhou, K. (2018). A cobalt-based metal–organic framework as cocatalyst on BiVO4 photoanode for enhanced photoelectrochemical water oxidation. *ChemSusChem* 11: 2710–2716.
[94] Shen, Q., Huang, X., Liu, J., Guo, C., Zhao, G. (2017). Biomimetic photoelectrocatalytic conversion of greenhouse gas carbon dioxide: Two-electron reduction for efficient formate production. *Applied Catalysis B: Environmental* 201: 70–76.
[95] Lo, -C.-C., Hung, C.-H., Yuan, C.-S., Wu, J.-F. (2007). Photoreduction of carbon dioxide with H_2 and H_2O over TiO_2 and ZrO_2 in a circulated photocatalytic reactor. *Solar Energy Materials and Solar Cells* 91: 1765–1774.
[96] Peng, Y.-P., Yeh, Y.-T., Shah, S.I., Huang, C.P. (2012). Concurrent photoelectrochemical reduction of CO_2 and oxidation of methyl orange using nitrogen-doped TiO_2. *Applied Catalysis B: Environmental* 123–124: 414–423.
[97] Zhang, Y., Xu, C., Chen, J., Zhang, X., Wang, Z., Zhou, J., Cen, K. (2015). A novel photo-thermochemical cycle for the dissociation of CO_2 using solar energy. *Applied Energy* 156: 223–229.
[98] Zhang, Y., Chen, J., Xu, C., Zhou, K., Wang, Z., Zhou, J., Cen, K. (2016). A novel photo-thermochemical cycle of water-splitting for hydrogen production based on TiO_{2-x}/TiO_2. *International Journal of Hydrogen Energy* 41: 2215–2221.
[99] Zhu, L., Gao, M., Nuo Peh, C.K., Wei Ho, G. (2018). Solar-driven photothermal nanostructured materials designs and prerequisites for evaporation and catalysis applications. *Materials Horizons* 5: 323–343.
[100] Ghoussoub, M., Xia, M., Duchesne, P.N., Segal, D., Ozin, G. (2019). Principles of photothermal gas-phase heterogeneous CO_2 catalysis. *Energy and Environmental Sciences*.
[101] Meng, X., Wang, T., Liu, L., Ouyang, S., Li, P., Hu, H., Kako, T., Iwai, H., Tanaka, A., Ye, J. (2014). Photothermal conversion of CO_2 into CH_4 with H_2 over group VIII nanocatalysts: an alternative approach for solar fuel production. *Angewandte Chemie International Edition* 53: 11478–11482.
[102] Ha, M.N., Lu, G., Liu, Z., Wang, L., Zhao, Z. (2016). 3DOM-$LaSrCoFeO_{6-\delta}$ as a highly active catalyst for the thermal and photothermal reduction of CO_2 with H_2O to CH_4. *Journal of Materials Chemistry A* 4: 13155–13165.
[103] Wang, L., Ha, M.N., Liu, Z., Zhao, Z. (2016). Mesoporous WO_3 modified by Mo for enhancing reduction of CO_2 to solar fuels under visible light and thermal conditions. *Integrated Ferroelectrics* 172: 97–108.

[104] Wang, L., Wang, Y., Cheng, Y., Liu, Z., Guo, Q., Ha, M.N., Zhao, Z. (2016). Hydrogen-treated mesoporous WO_3 as a reducing agent of CO_2 to fuels (CH_4 and CH_3OH) with enhanced photothermal catalytic performance. *Journal of Materials Chemistry A* 4: 5314–5322.
[105] Li, Y., Wang, C., Song, M., Li, D., Zhang, X., Liu, Y. (2019). TiO_{2-x}/CoO_x photocatalyst sparkles in photothermocatalytic reduction of CO_2 with H_2O steam. *Applied Catalysis B: Environmental* 243: 760–770.
[106] Yang, J., Guo, Y., Lu, W., Jiang, R., Wang, J. (2018). Emerging applications of plasmons in driving CO_2 reduction and N_2 fixation. *Advanced Materials* 30: 1802227.
[107] Li, J., Ye, Y., Ye, L., Su, F., Ma, Z., Huang, J., Xie, H., Doronkin, D.E., Zimina, A., Grunwaldt, J.-D., Zhou, Y. (2019). Sunlight induced photo-thermal synergistic catalytic CO_2 conversion via localized surface plasmon resonance of MoO_{3-x}. *Journal of Materials Chemistry A* 7: 2821–2830.
[108] Zhang, H., Wang, T., Wang, J., Liu, H., Dao, T.D., Li, M., Liu, G., Meng, X., Chang, K., Shi, L., Nagao, T., Ye, J. (2016). Surface-plasmon-enhanced photodriven CO_2 reduction catalyzed by metal–organic-framework-derived iron nanoparticles encapsulated by ultrathin carbon layers. *Advanced Materials* 28: 3703–3710.
[109] Low, J., Zhang, L., Zhu, B., Liu, Z., Yu, J. (2018). TiO_2 photonic crystals with localized surface photothermal effect and enhanced photocatalytic CO_2 reduction activity. *ACS Sustainable Chemistry & Engineering* 6: 15653–15661.
[110] Hoch, L.B., O'Brien, P.G., Jelle, A., Sandhel, A., Perovic, D.D., Mims, C.A., Ozin, G.A. (2016). Nanostructured indium oxide coated silicon nanowire arrays: A hybrid photothermal/photochemical approach to solar fuels. *ACS Nano* 10: 9017–9025.
[111] Chen, G., Gao, R., Zhao, Y., Li, Z., Waterhouse, G.I.N., Shi, R., Zhao, J., Zhang, M., Shang, L., Sheng, G., Zhang, X., Wen, X., Wu, L.-Z., Tung, C.-H., Zhang, T. (2018). Alumina-supported cofe alloy catalysts derived from layered-double-hydroxide nanosheets for efficient photothermal CO_2 hydrogenation to hydrocarbons. *Advanced Materials* 30: 1704663.
[112] An, W.-J., Thimsen, E., Biswas, P. (2010). Aerosol-chemical vapor deposition method for synthesis of nanostructured metal oxide thin films with controlled morphology. The *Journal of Physical Chemistry Letters* 1: 249–253.
[113] Wang, W.-N., An, W.-J., Ramalingam, B., Mukherjee, S., Niedzwiedzki, D.M., Gangopadhyay, S., Biswas, P. (2012). Size and structure matter: Enhanced CO_2 photoreduction efficiency by size-resolved ultrafine Pt nanoparticles on TiO_2 single crystals. *Journal of the American Chemical Society* 134: 11276–11281.
[114] Guo, J., Yang, Z., Archer, L.A. (2013). Aerosol assisted synthesis of hierarchical tin–carbon composites and their application as lithium battery anode materials. *Journal of Materials Chemistry A* 1: 8710–8715.
[115] Yu, H., Shi, R., Zhao, Y., Waterhouse, G.I.N., Wu, L.-Z., Tung, C.-H., Zhang, T. (2016). Smart utilization of carbon dots in semiconductor photocatalysis. *Advanced Materials* 28: 9454–9477.
[116] Hou, J., Cheng, H., Yang, C., Takeda, O., Zhu, H. (2015). Hierarchical carbon quantum dots/hydrogenated-γ-TaON heterojunctions for broad spectrum photocatalytic performance. *Nano Energy* 18: 143–153.
[117] Motl, N.E., Mann, A.K.P., Skrabalak, S.E. (2013). Aerosol-assisted synthesis and assembly of nanoscale building blocks. *Journal of Materials Chemistry A* 1: 5193–5202.
[118] Wang, W.-N., Park, J., Biswas, P. (2011). Rapid synthesis of nanostructured $Cu–TiO_2–SiO_2$ composites for CO_2 photoreduction by evaporation driven self-assembly. *Catalysis Science and Technology* 1: 593–600.

[119] Zhan, Z., Wang, W.-N., Zhu, L., An, W.-J., Biswas, P. (2010). Flame aerosol reactor synthesis of nanostructured SnO_2 thin films: High gas-sensing properties by control of morphology. *Sensors and Actuators. B Chemical* 150: 609–615.

[120] Tiwari, V., Jiang, J., Sethi, V., Biswas, P. (2008). One-step synthesis of noble metal–titanium dioxide nanocomposites in a flame aerosol reactor. *Applied Catalysis A: General* 345: 241–246.

[121] Abram, C., Shan, J., Yang, X., Yan, C., Steingart, D., Ju, Y. (2019). Flame aerosol synthesis and electrochemical characterization of ni-rich layered cathode materials for li-ion batteries. *ACS Applied Energy Materials* 2: 1319–1329.

[122] Strobel, R., Pratsinis, S.E. (2007). Flame aerosol synthesis of smart nanostructured materials. *Journal of Materials Chemistry* 17: 4743–4756.

[123] Koirala, R., Pratsinis, S.E., Baiker, A. (2016). Synthesis of catalytic materials in flames: Opportunities and challenges. *Chemical Society Reviews* 45: 3053–3068.

[124] Teoh, W.Y., Amal, R., Mädler, L. (2010). Flame spray pyrolysis: An enabling technology for nanoparticles design and fabrication. *Nanoscale* 2: 1324–1347.

[125] Tada, S., Larmier, K., Büchel, R., Copéret, C. (2018). Methanol synthesis via CO_2 hydrogenation over CuO–ZrO_2 prepared by two-nozzle flame spray pyrolysis. *Catalysis Science and Technology* 8: 2056–2060.

[126] Schubert, M., Pokhrel, S., Thomé, A., Zielasek, V., Gesing, T.M., Roessner, F., Mädler, L., Bäumer, M. (2016). Highly active Co–Al_2O_3 -based catalysts for CO_2 methanation with very low platinum promotion prepared by double flame spray pyrolysis. *Catalysis Science and Technology* 6: 7449–7460.

Ramesh Raliya and Katie Halwachs

Chapter 6 Aerosol science and nanoscale engineering enabling agriculture

Abstract: Increasing use of agrochemicals to improve crop production is a growing concern for both farmers and consumers. The major challenges are associated with the fertilizer uptake rate of nitrogen and phosphorus granular fertilizers, which is less than 30%. The unabsorbed fertilizer runs off into freshwater bodies and causes eutrophication. To address this problem, scientists are exploring nanoscale nutrients to be utilized as plant fertilizers. The initial trend shows that the nanoparticles composed of metal-, metal oxide-, carbon-, and polymer-based composites have the ability to improve crop productivity while minimizing the loss of nutrients. The aforementioned nanoparticles were synthesized utilizing top-down and bottom-up fabrication techniques such as aerosol methods. The aerosol-based nanoparticle synthesis enables the scaling up of manufacturing with more controlled physicochemical properties. In addition, aerosol-based foliar applications further assist in gas-phase nanoparticle uptake through plants' stomatal openings. This chapter explains the importance of aerosol science and nanoscale engineering application in agriculture.

Keywords: aerosol, nanoparticles, foliar delivery, agrochemicals, nutrient uptake

6.1 Introduction

Aerosols are miniscule particles of solids or droplets of liquid that are dispersed in gas. The size of particles ranges from 0.001 to 100 μm, typically polydisperse in size and composition [1–3]. These suspensions of particles are encountered on a daily basis, can be natural or manufactured, and are commonly found in smoke, pollution,

Acknowledgments: The authors are grateful to the LEAP Inventor Challenge Award granted by the Skandalaris Center for Interdisciplinary Innovation and Entrepreneurship at Washington University in St. Louis. We also thank Prof. Pratim Biswas, Aerosol and Air Quality Research Laboratory, Washington University, for the necessary resources.

Ramesh Raliya, Department of Energy, Environmental and Chemical Engineering, Washington University in Saint Louis, MO 63130, USA, e-mail: rameshraliya@iffco.in; IFFCO – Nano Biotechnology Research Center, Gandhinagar 382423, Gujarat, India
Katie Halwachs, Department of Energy, Environmental and Chemical Engineering, Washington University in Saint Louis, MO 63130, USA, e-mail: rameshraliya@iffco.in

https://doi.org/10.1515/9783110729481-006

spray products, medical applications, and, most importantly, agriculture. Based on the type and impact, aerosols can be divided into two groups: good aerosols, or those having useful applications, such as engineered nanoparticle-based aerosol coatings, and bad aerosols, such as aerosols contained in vehicular emissions, which affect human health.

The traditional use of aerosols is a fine spray. For example, household uses of these products include hairspray, deodorant, sunscreen, furniture polish, and air fresheners. Medical applications include nasal sprays and oral inhalers. While these applications are all useful, some aerosols, often considered pollution, are harmful to biological life. For example, toxic particulate matter from the combustion of fuel, burning of biomass, and noncombustion sources like road dust or ammonium nitrates lead to adverse health effects including respiratory disease, lung cancer, and premature mortality. In addition, atmospheric aerosols cause harm to plants by affecting their growth conditions including temperature, amount of radiation they receive, and quality and amount of water and nutrients. The main consequence of aerosols on plants is that they absorb light as it goes through the atmosphere, decreasing the amount received by plants, lowering their photosynthetic activity, and possibly causing a significant decrease on crop yields [4].

Atmospheric aerosol is only one challenge that the agricultural sector is facing today. Another major problem involves the low-efficacy agricultural aerosols being utilized. Many products are aerosolized and applied to the crops including pesticides, herbicides, antifungal and antibacterial treatments, fertilizers, and simple water jets. With the use of these items comes a plethora of problems. For example, runoff degrades land and causes eutrophication of waterways, poor efficiency of nutrient utilization, decreasing quantities of cultivatable land, and depletion of a limited water supply [5]. Moreover, food demand is increasing at an unprecedented rate, the population is expected to reach over 9 billion by 2050, and stores of natural resources are steadily decreasing. All of these factors combine to create quite the quandary and will require major innovation to reach a solution. In this chapter, we review the recent research and attempts that are in progress to expand the use of (good) aerosols for sustainable and precision agriculture while utilizing nanotechnology. Nanotechnology, a modern field of the latest century, is the subject that seeks to improve the lives of humans and involves materials, chemicals, and organic matter with unique physicochemical properties. It involves engineered objects with size ranges between 1 and 100 nm at least at one dimension. Aerosol technologies and concepts are being tested and utilized for both the manufacturing and application of nanoscale materials.

6.2 Aerosol-based nanoparticle synthesis

The general process of aerosol manufacturing starts with the preparation of raw precursor materials by mining and refining ores or by using recycled gases and liquid-phase molecular precursors. The desired molecular precursor is then passed through flame [6], plasma [7], hot air furnace [8, 9], vapor deposition, or a combination of these reactors [10, 11]. The aerosol reactors transform these raw or molecular state precursor materials into nanomaterials through aerosol processes driven by high temperatures. The high-temperature exposure is followed by cooling to stop nanoparticle growth by the collisional and condensational mechanism. Depending on the desired product's properties, particles are collected from the gas phase immediately or are conditioned by aerosol coating and/or functionalization steps. Major advantages of the aerosol methods for nanoparticles synthesis include (a) low cost, scalable method of manufacturing for sophisticated nanoscale particles and assembly or coating of such nanomaterial into functional devices; (b) aerosol processes do not generate liquid byproducts, which are costly and difficult to dispose of. This is favored by industry because it reduces water consumption and the cost of waste disposal treatment. (c) This is a gas-phase synthesis; therefore, highly pure materials can be synthesized using flame and furnace reactors. (d) Compared to other processes, aerosol methods require simpler and fewer operational assembly units. (e) Novel materials can be synthesized by rapid phase change. (f) Capture, collection, and characterization of particles are easier and cheaper than wet chemical methods. (g) Particle sintering, diffusion, and transport are more efficient than other nanomaterial synthesis approaches. (h) It is easy to make stable composite nanostructure despite relatively low material compatibility. All of these advantages lead to significant development in the scaling up of nanomaterial manufacturing and application thereof [12, 13]. Using the aerosol methods, nanoparticles of titania, silica, carbon black, and alumina are being manufactured at tons per hour for commercial applications including paint, medicines, and construction materials. While there are many advantages to the aerosol-based synthesis methods, there are some disadvantages. The major challenges of aerosol-based nanoparticle synthesis are the aggregation of primary particles due to high-temperature gas-phase processes, the expensive precursor chemicals, and the nonhomogeneous morphology and particle size distribution [14]. Strategies to address these complications can eventually be found using the art of engineering novel methods for aerosol synthesis of nanomaterials.

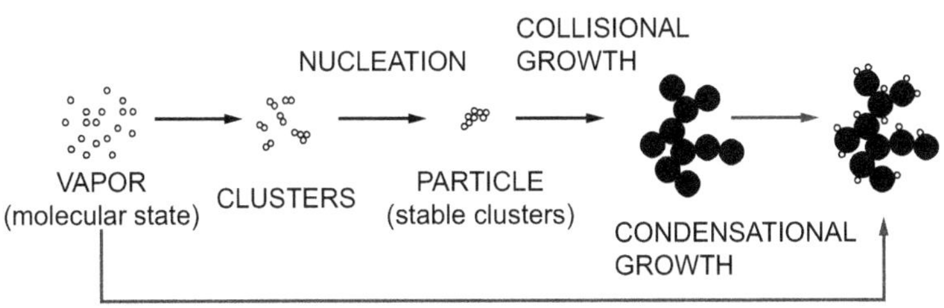

Figure 6.1: Basic steps of nanoparticle synthesis (bottom-up fabrication) using the aerosol method.

6.3 Nanoparticles used for agricultural application

Various nanoscale materials such as inorganic materials, organic matter, and composites of metals or nonmetals are having their performance assessed, and their impact on plant growth, development, and productivity is being investigated. A wide range of nanoparticles are being used as nanofertilizers and their progress can be read about in a recent review article summarizing nanofertilizers for precision and sustainable agriculture [5]. There are a variety of mechanisms for applying the nanoscale particles that are having their agricultural benefits explored. Nanoscale particles in agriculture can be delivered by seed coating, soil amendment, and foliar aerosol spray. The nanoparticles that can be delivered through foliar aerosol spray are listed and their impact on various crops is summarized in Table 6.1.

Table 6.1: Nanoparticles that can be delivered through foliar aerosol spray and influence plant growth.

Type of nanoparticles (NPs)	NP properties		NP exposure to plants	Tested plant	Major observation
	Size (nm)	Concentration (ppm)			
Essential plant nutrient					
Carbon Carbon-based NPs (CNT, graphene) MWCNT; SWCNT; fullerol	1.5–5	5–500	Nutrient media and **foliar uptake**; seed treatment	Tomato [15–17], tobacco [16, 18, 19] Wheat [20], gram [21, 22] Bitter melon [23] Saltmarsh cordgrass [24] soybean [17, 25, 26] Corn, barley, rice, switchgrass [17, 26]	– Promote upregulation of stress-related genes; promote in vitro growth and biomass – Enhanced root elongation – Improve crop yield and seed quality – Reduce heavy metal toxicity and stress
Phosphorus CaPo4, CMC – HA, Phosphorite Zn-induced P	<50	10–100	Soil and **foliar applications**	Cotton [27], pearl millets [28] Beans [29–31] Wheat, rye, pea, barley, corn, buckwheat, radish, cucumber [32]	– Protect against oxidative stress – Mobilize native P and enhance uptake – Enhance plant growth and yield

Table 6.1 (continued)

Type of nanoparticles (NPs)	NP properties		NP exposure to plants	Tested plant	Major observation
	Size (nm)	Concentration (ppm)			
Magnesium MgO	<10	15	**Foliar**	Clusterbean [33]	– Improve biomass, chlorophyll content, and phenological growth
Copper Cu-chitosan	>10	100–1,200	**Foliar** and seed treatment	Corn [34], tomato [35]	– Enhance seedling growth, plant biomass, and biochemical activities
Zinc ZnO	20–30	10–2,000	**Foliar** application Seed application	Peanut [36] Beans [29, 30, 37, 38] Tomato [39] Cotton [27] Maize [40]	– Increase yield potential and plant growth – Enhance phytohormone level and plant growth – Help reduce drought stress and improve agronomic fortification
Iron Iron oxide	10–100	1.5–4,000	**Foliar** spray	Wheat [41], watermelon [42, 43] Clover [44], soybean [45–47] Rice [48], tomato [49] Peanut [50], corn [51] Pumpkin [52]	– Enhance photosynthesis rate, chlorophyll content, biomass, grain yield, and nutritional quality – Improve plant growth – Enhance nutrient absorption by enhancing microbial enzyme activity in rhizosphere

Table 6.1 (continued)

Type of nanoparticles (NPs)	NP properties		NP exposure to plants	Tested plant	Major observation
	Size (nm)	Concentration (ppm)			
Nonessential plant nutrient					
Titanium TiO_2	5–100	200–600	Seed, soil, and **foliar** exposure	Spinach [53–61] *Lemna minor* [62] Tomato [39, 49], wheat [63] Watermelon [42], mung bean [64], moth bean [65] Pearl millet [65] Clusterbean [65, 66]	– Increase plant biomass and photosynthetic activity – Enhance biochemical enzyme activity and light absorption by chloroplast – Increase photosynthesis, RuBISCO activity, and carbon fixation – Increase germination rate – Enhance nitrogen metabolism

Aerosolizing nanoparticles and applying them to plants via foliar application can provide controlled release, targeted delivery, enhanced leaf adhesion, and increased efficacy, all of which allow for reduced rates of application, maximize yields, and minimize negative agricultural impacts [67]. These advantages of aerosolized nanoparticle agriculture products over regular agricultural products stem from the unique characteristics of nanoparticles. Some of these characteristics include the adeptness of nanoparticles in promoting electron exchange, maximized surface area, and enhanced surface reactive abilities [68]. One of the biggest challenges in the efficacy of agricultural products delivered via foliar application is the absorption and translocation in the plant. When choosing the size of the nanoparticles, the size of the pore sizes of plant cell walls should be heavily considered [68]. Through synthesis, the size of nanoparticles can be tailored for maximal uptake in specific plant species, and they can be homogeneously produced through modern methods.

Aerosol applications ensure that the nanoparticles sustain an adequate size for uptake and boost the uptake of particles by bypassing the main barrier of plant tissues, also known as the cuticle [68]. In addition, aerosolized particles are monodisperse and are somewhat more stable than particles in regular suspensions, which coagulate from the interactions between particles [42, 69]. Currently, five subcategories of nanoparticles are being developed for advancing agrochemical products. These are metal nanoparticles, metal oxide nanoparticles, carbon-based nanomaterials, organic polymer, and composites of two or more of the previously mentioned subcategories. All of these particles can be aerosolized and applied to the plants via foliar application, and they serve a variety of functions from uptake to metabolism.

Metal nanoparticles have unique features such as generating oxidative radicals in the cellular system, helping avoid pathogenic attack by mimicking antimicrobial properties, acting as cofactors for various biomolecules, and regulating metabolomics. Furthermore, metal nanoparticles with enhanced surface plasmon response are being used to develop sensors for monitoring biochemical pathways or detecting products in the plant system [70]. Metals serve particular purposes in cellular function that cannot be filled by organic molecules, making them necessary for life; however, these metals become lethal for plant itself when present in excess [67]. Mechanisms for which metal nanoparticles serve as antimicrobial agents involve causing injuries to microbial cells through stress, functional disruption, or membrane damage. An example of one of these mechanisms can be found in silver nanoparticles, which can be used in agriculture via a foliar, nanoparticle application. Silver nanoparticles function as antimicrobial agents by releasing ions that interact with thiol groups in proteins, negatively influencing DNA replication, and eventually leading to cell death [71].

Nanoparticles of metal oxides have many applications, but the most prevalent uses in agriculture are as agrochemicals, most notably as nanofertilizers or nanopesticides [5, 67]. Some of the most common metal oxides used are zinc oxide (ZnO), iron oxide (Fe_2O_3), silicon dioxide (SiO_2), and titanium dioxide (TiO_2) or their composites with other essential micro- and macronutrients. Zinc oxide nanoparticles enhance Zn biofortification in the plant; can increase the absorption of other native nutrients from the soil, such as phosphorus; increase nutritional quality of crops; and improve the crop yield when applied as fertilizers [5]. Similarly, iron oxide has been shown to improve traits such as biomass, crop yield, nutritional quality, chlorophyll content, photosynthetic activity, and absorption of phosphorus and nitrogen. Silicon oxide has comparable effects to zinc oxide and iron oxide, as well as stimulating seedling growth, increasing stress tolerance of the plants, and increasing the level of leaf water content. Although titanium dioxide nanoparticles have similar effects to iron oxide, with the addition of an increased germination rate, it is not an essential plant nutrient element like iron oxide. The effects of metal oxides used as fertilizers offer a solution to the current issue of marginal uptake of fertilizers which creates a high demand of fertilizers to maintain crop productivity. This large volume of applied fertilizer that runs off into waterways creates environmental problems such as eutrophication and greenhouse gas emissions.

Carbon nanomaterials such as graphene, fullerene, and carbon nanotube are being used as a carbon source to grow the plants and map their molecular interactions. Carbon is a major constituent for biomolecules/biopolymers such as proteins, carbohydrates, and lipids. Plants also use carbon in the form of carbon dioxide, which they uptake during photosynthesis. Studies utilizing carbon nanotubes as fertilizers have reported increased growth of roots and shoots, an improved germination rate, higher nutrient uptake, and greater fresh biomass [72]. Furthermore, due to the carbon nanotubules' tightly arranged atoms linked through sp^2 bonding, they are extremely strong fibers and possess an exceptional mix of transduction, magnetic, optical, and chemical properties that make them remarkably astute sensors [73]. Their success as sensors can also be attributed to their extremely large specific surface area, which allows for a large number of receptors to be attached, their ability to be opened and filled with materials without sacrificing functionality or stability, and the ability of altered carbon nanotubules to successfully transverse biological barriers [73].

Similarly, other nanoparticles such as liposomes, which are vesicles with a central aqueous core that are encapsulated by a lipid bilayer, are stable in aqueous solutions and readily fuse to the plasma membrane of cells or are consumed by cells through endocytosis [74]. In the agricultural sector, liposomes are most frequently utilized for delivering active ingredients to plants and as detection sensors. Using liposomes to administer active ingredients via foliar application has proven quite effective, demonstrating a 33-fold increase in leaf penetration compared to free molecules employed in a similar manner. This increase is due to the ability of liposomes to penetrate the leaves and move through the plant's vascular system without resulting in toxicity in the plant. Liposomes are used in farming as nanosensors for pathogen detection, pesticide detection, and toxin detection. For pathogen detection, immunoliposomes are created by tagging liposomes with antibodies against the target pathogen and then detecting them using fluorescent signals [74]. Immunoliposome assays are also utilized in pesticide detection, and liposomes with toxin receptors on the surface are utilized in toxin detection. Utilizing liposomes in this way provides a great advantage over traditional sensors because of their affordability, biocompatibility, biodegradability, readily modified surface, and sensitivity in immunoassays.

6.4 Aerosol-based nanoparticle delivery to plants

Foliar spray application of agrochemicals (nutrients and pesticides) has been practiced for years [75]. In general, fertilization by soil amendment or seed coating with nutrients is practiced on the basis of nutrient deficiency in the soil, whereas aerosol-based foliar application is done to alleviate nutrient deficiency symptoms exhibited by plants [76]. It is not explained clearly, however, that the deficiency is correlated with the soil and how the long-term soil nutrient mining by plants impacts the soil biological health. Foliar application is most effective when a plant has a high leaf area index and requires low exposure dosage and minimized application replicates. However, timing of the foliar aerosol application is dependent on environmental conditions such as temperature, air-wind flow, humidity, and sunlight. To avoid nutrient loss and impact on other organisms, including inhalation toxicity to human and other animals, the time of application must be optimized [77]. Furthermore, the toxicity further depends on particle concentration in the atmosphere and physicochemical properties of the particles [78–83]. Therefore, it is recommended to wear appropriate personal protection equipment such as a mask, gloves, and eye protection when applying nanoscale fertilizers through foliar aerosol spray [84–86]. The major advantage of aerosol application is the ability to maintain the effective droplet size of the agrochemical or target compound containing particles. By using a fine aerosol generator, the user can spray monodisperse droplets, while conventional suspension spray or soil applications undergo agglomeration as a result of particle–particle or particle–soil interactions [42, 69]. Nanoscale material-based agrochemical properties are particularly important for foliar delivery, where size exclusion from stomatal openings may limit the uptake of gas-phase particles [87]. It has been demonstrated that stomatal uptake is enhanced through control of nanoscale materials' particle size combined with an aerosol delivery method [69]. This enhancement can be observed in the aerosol-based foliar delivery of iron and magnesium nanoparticles to plants, which improves biomass or foliage growth [8]. More specifically, aerosol-mediated nanoparticle delivery, penetration, and translocation have been studied in depth in watermelon [42, 69] and tomato [89] plants [5].

6.5 Summary and future perspectives

In summary, aerosol science and engineering provides a potential solution for the problems associated with agrochemicals, particularly fertilizers that cause environmental pollution. Gas-phase synthesis mechanisms manufacture nanoscale particles that can be used for plant nutrition without generating liquid waste byproducts. Moreover, aerosol spray of manufactured nanoparticles on plant leaves further improves their uptake rate. Because of nanoparticles' higher surface area to volume ratio and quicker uptake rate stemming from tiny particle size and fine particle spray, aerosol applications of nanomaterials have the potential to reduce fertilizer demand by means of mass. However, engineered nanomaterials may cause toxicity if inhaled beyond the critical particle concentration. Therefore, it is recommended to use appropriate safety precautions for the application of nanoscale materials for agricultural uses. Despite many advantages of the aerosol method, aggregation or agglomeration of the particles, polydispersity of the particles, and higher cost of molecular precursor compared to wet chemistry largely limit the widespread application in agriculture. In closing, nanotechnology-enabled aerosol science paves new ways of manufacturing agrochemicals and optimizing their delivery to plants to minimize environmental pollution and boost precision and sustainability.

References

[1] Peretz, D.H. (2009). Aerosols: Chemistry, Environmental Impact and Health Effects, Nova Science Publ.

[2] Kulkarni, P., Baron, P.A., Willeke, K. (2011). Aerosol Measurement: Principles, Techniques, and Applications, John Wiley & Sons.

[3] Hinds, W.C. (2012). Aerosol Technology: Properties, Behavior, and Measurement of Airborne Particles, John Wiley & Sons.

[4] Greenwald, R., Bergin, M., Xu, J., Cohan, D., Hoogenboom, G., Chameides, W. (2006). The influence of aerosols on crop production: A study using the CERES crop model. *Agricultural Systems* 89(2–3): 390–413.

[5] Raliya, R., Saharan, V., Dimkpa, C., Biswas, P. (2018). Nanofertilizer for precision and sustainable agriculture: Current state and future perspectives. *Journal of Agricultural and Food Chemistry* 66(26): 6487–6503.

[6] Li, S., Ren, Y., Biswas, P., Stephen, D.T. (2016). Flame aerosol synthesis of nanostructured materials and functional devices: Processing, modeling, and diagnostics. *Progress in Energy and Combustion Science* 55: 1–59.

[7] Mangolini, L., Thimsen, E., Kortshagen, U. (2005). High-yield plasma synthesis of luminescent silicon nanocrystals. *Nano Letters* 5(4): 655–659.

[8] Fortner, J., Biswas, P., Jiang, Y., Wang, W. (2019). Composite nanostructures having a crumpled graphene oxide shell. Google Patents.

[9] Jiang, Y., Raliya, R., Liao, P., Biswas, P., Fortner, J.D. (2017). Graphene oxides in water: Assessing stability as a function of material and natural organic matter properties. *Environmental Science: Nano* 4(7): 1484–1493.

[10] Chadha, T.S. (2016). Nanostructured Thin Film Synthesis by Aerosol Chemical Vapor Deposition for Energy Storage Applications.

[11] Chadha, T.S., Dutta, P.K., Raliya, R., Mitra, S., Biswas, P. (2018). Sustainable one step process for making carbon-free TiO_2 anodes and sodium-ion battery electrochemistry. *Sustainable Energy & Fuels* 2(7): 1582–1587.

[12] Buesser, B., Pratsinis, S.E. (2012). Design of nanomaterial synthesis by aerosol processes. *Annual Review of Chemical and Biomolecular Engineering* 3: 103–127.

[13] Flagan, R.C. (1998). Nanoparticles and nanostructures: Aerosol synthesis and characterization. In: *Nanostructured Materials*, Springer, 15–30.

[14] Eggersdorfer, M.L., Pratsinis, S.E. (2014). Agglomerates and aggregates of nanoparticles made in the gas phase. *Advanced Powder Technology* 25(1): 71–90.

[15] Khodakovskaya, M.V., Kim, B.S., Kim, J.N., Alimohammadi, M., Dervishi, E., Mustafa, T., Cernigla, C.E. (2013). Carbon nanotubes as plant growth regulators: Effects on tomato growth, reproductive system, and soil microbial community. *Small* 9(1): 115–123.

[16] Lahiani, M.H., Dervishi, E., Ivanov, I., Chen, J., Khodakovskaya, M. (2016). Comparative study of plant responses to carbon-based nanomaterials with different morphologies. *Nanotechnology* 27(26): 265102.

[17] Lahiani, M.H., Chen, J., Irin, F., Puretzky, A.A., Green, M.J., Khodakovskaya, M.V. (2015). Interaction of carbon nanohorns with plants: Uptake and biological effects. *Carbon* 81: 607–619.
[18] Khodakovskaya, M.V., de Silva, K., Biris, A.S., Dervishi, E., Villagarcia, H. (2012). Carbon nanotubes induce growth enhancement of tobacco cells. *ACS Nano* 6(3): 2128–2135.
[19] Villagarcia, H., Dervishi, E., de Silva, K., Biris, A.S., Khodakovskaya, M.V. (2012). Surface chemistry of carbon nanotubes impacts the growth and expression of water channel protein in tomato plants. *Small* 8(15): 2328–2334.
[20] Wang, X., Han, H., Liu, X., Gu, X., Chen, K., Lu, D. (2012). Multi-walled carbon nanotubes can enhance root elongation of wheat (*Triticum aestivum*) plants. *Journal of Nanoparticle Research* 14(6): 841.
[21] Sonkar, S.K., Roy, M., Babar, D.G., Sarkar, S. (2012). Water soluble carbon nano-onions from wood wool as growth promoters for gram plants. *Nanoscale* 4(24): 7670–7675.
[22] Tripathi, S., Sonkar, S.K., Sarkar, S. (2011). Growth stimulation of gram (*Cicer arietinum*) plant by water soluble carbon nanotubes. *Nanoscale* 3(3): 1176–1181.
[23] Kole, C., Kole, P., Randunu, K.M., Choudhary, P., Podila, R., Ke, P.C., Rao, A.M., Marcus, R.K. (2013). Nanobiotechnology can boost crop production and quality: First evidence from increased plant biomass, fruit yield and phytomedicine content in bitter melon (*Momordica charantia*). *BMC Biotechnology* 13(1): 37.
[24] Chai, M., Shi, F., Li, R., Liu, L., Liu, Y., Liu, F. (2013). Interactive effects of cadmium and carbon nanotubes on the growth and metal accumulation in a halophyte Spartina alterniflora (Poaceae). *Plant Growth Regulation* 71(2): 171–179.
[25] Wand, X., Huang, Q., Wang, L., Wang, L. (2012). Effect of single-wall carbon nanotube on soybean (*Glycine max*) regeneration from mature cotyledonary node explants. *Nano Life* 2 (04): 1250014.
[26] Lahiani, M.H., Dervishi, E., Chen, J., Nima, Z., Gaume, A., Biris, A.S., Khodakovskaya, M.V. (2013). Impact of carbon nanotube exposure to seeds of valuable crops. *ACS Applied Materials & Interfaces* 5(16): 7965–7973.
[27] Venkatachalam, P., Priyanka, N., Manikandan, K., Ganeshbabu, I., Indiraarulselvi, P., Geetha, N., Muralikrishna, K., Bhattacharya, R.C., Tiwari, M., Sharma, N., Sahi, S.V. (2017). Enhanced plant growth promoting role of phycomolecules coated zinc oxide nanoparticles with P supplementation in cotton (*Gossypium hirsutum* L.). *Plant Physiology and Biochemistry* 110: 118–127.
[28] Tarafdar, J., Raliya, R., Mahawar, H., Rathore, I. (2014). Development of zinc nanofertilizer to enhance crop production in pearl millet (*Pennisetum americanum*). *Agricultural Research* 3 (3): 257–262.
[29] Raliya, R., Tarafdar, J.C. (2013). ZnO nanoparticle biosynthesis and its effect on phosphorous-mobilizing enzyme secretion and gum contents in clusterbean (*Cyamopsis tetragonoloba* L.). *Agricultural Research* 2(1): 48–57.
[30] Raliya, R., Tarafdar, J.C., Biswas, P. (2016). Enhancing the mobilization of native phosphorus in the mung bean rhizosphere using ZnO nanoparticles synthesized by soil fungi. *Journal of Agricultural and Food Chemistry* 64(16): 3111–3118.
[31] Liu, R., Lal, R. (2014). Synthetic apatite nanoparticles as a phosphorus fertilizer for soybean (Glycine max). *Scientific Reports* 4: 5686.
[32] Sharonova, N.L., Yapparov, A.K., Khisamutdinov, N.S., Ezhkova, A.M., Yapparov, I.A., Ezhkov, V.O., Degtyareva, I.A., Babynin, E.V. (2015). Nanostructured water-phosphorite suspension is a new promising fertilizer. *Nanotechnologies in Russia* 10(7): 651–661.
[33] Raliya, R., Tarafdar, J., Singh, S., Gautam, R., Choudhary, K., Maurino, V.G., Saharan, V. (2014). MgO nanoparticles biosynthesis and its effect on chlorophyll contents in the leaves

of clusterbean (*Cyamopsis tetragonoloba* L.). *Advanced Science, Engineering and Medicine* 6 (5): 538–545.
[34] Saharan, V., Kumaraswamy, R., Choudhary, R.C., Kumari, S., Pal, A., Raliya, R., Biswas, P. (2016). Cu-chitosan nanoparticle mediated sustainable approach to enhance seedling growth in maize by mobilizing reserved food. *Journal of Agricultural and Food Chemistry* 64 (31): 6148–6155.
[35] Saharan, V., Sharma, G., Yadav, M., Choudhary, M.K., Sharma, S., Pal, A., Raliya, R., Biswas, P. (2015). Synthesis and in vitro antifungal efficacy of Cu–chitosan nanoparticles against pathogenic fungi of tomato. *International Journal of Biological Macromolecules* 75: 346–353.
[36] Prasad, T., Sudhakar, P., Sreenivasulu, Y., Latha, P., Munaswamy, V., Reddy, K.R., Sreeprasad, T., Sajanlal, P., Pradeep, T. (2012). Effect of nanoscale zinc oxide particles on the germination, growth and yield of peanut. *Journal of Plant Nutrition* 35(6): 905–927.
[37] Pandey, A.C., S. Sanjay, S., S. Yadav, R. (2010). Application of ZnO nanoparticles in influencing the growth rate of *Cicer arietinum*. *Journal of Experimental Nanoscience* 5(6): 488–497.
[38] Dimkpa, C.O., Bindraban, P.S., Fugice, J., Agyin-Birikorang, S., Singh, U., Hellums, D. (2017). Composite micronutrient nanoparticles and salts decrease drought stress in soybean. *Agronomy for Sustainable Development* 37(1): 5.
[39] Raliya, R., Nair, R., Chavalmane, S., Wang, W.-N., Biswas, P. (2015). Mechanistic evaluation of translocation and physiological impact of titanium dioxide and zinc oxide nanoparticles on the tomato (*Solanum lycopersicum* L.) plant. *Metallomics* 7(12): 1584–1594.
[40] Subbaiah, L.V., Prasad, T.N.V.K.V., Krishna, T.G., Sudhakar, P., Reddy, B.R., Pradeep, T. (2016). Novel effects of nanoparticulate delivery of zinc on growth, productivity, and zinc biofortification in maize (*Zea mays* L.). *Journal of Agricultural and Food Chemistry* 64(19): 3778–3788.
[41] Ghafari, H., Razmjoo, J. (2013). Effect of foliar application of nano-iron oxidase, iron chelate and iron sulphate rates on yield and quality of wheat. *International Journal of Agronomy and Plant Production* 4(11): 2997–3003.
[42] Wang, W.-N., Tarafdar, J.C., Biswas, P. (2013). Nanoparticle synthesis and delivery by an aerosol route for watermelon plant foliar uptake. *Journal of Nanoparticle Research* 15(1): 1–13.
[43] Li, J., Chang, P.R., Huang, J., Wang, Y., Yuan, H., Ren, H. (2013). Physiological effects of magnetic iron oxide nanoparticles towards watermelon. *Journal of Nanoscience and Nanotechnology* 13(8): 5561–5567.
[44] Feng, Y., Cui, X., He, S., Dong, G., Chen, M., Wang, J., Lin, X. (2013). The role of metal nanoparticles in influencing arbuscular mycorrhizal fungi effects on plant growth. *Environmental Science & Technology* 47(16): 9496–9504.
[45] Ghafariyan, M.H., Malakouti, M.J., Dadpour, M.R., Stroeve, P., Mahmoudi, M. (2013). Effects of magnetite nanoparticles on soybean chlorophyll. *Environmental Science & Technology* 47 (18): 10645–10652.
[46] Alidoust, D., Isoda, A. (2013). Effect of γFe_2O_3 nanoparticles on photosynthetic characteristic of soybean (Glycine max (L.) Merr.): Foliar spray versus soil amendment. *Acta Physiologiae Plantarum* 35(12): 3365–3375.
[47] Sheykhbaglou, R., Sedghi, M., Shishevan, M.T., Sharifi, R.S. (2010). Effects of nano-iron oxide particles on agronomic traits of soybean. *Notulae Scientia Biologicae* 2(2): 112.
[48] Alidoust, D., Isoda, A. (2014). Phytotoxicity assessment of γ-Fe_2O_3 nanoparticles on root elongation and growth of rice plant. *Environmental Earth Sciences* 71(12): 5173–5182.
[49] Giordani, T., Fabrizi, A., Guidi, L., Natali, L., Giunti, G., Ravasi, F., Cavallini, A., Pardossi, A. (2012). Response of tomato plants exposed to treatment with nanoparticles. *EQA-International Journal of Environmental Quality* 8(8): 27–38.

[50] Rui, M., Ma, C., Hao, Y., Guo, J., Rui, Y., Tang, X., Zhao, Q., Fan, X., Zhang, Z., Hou, T. (2016). Iron oxide nanoparticles as a potential iron fertilizer for peanut (*Arachis hypogaea*). *Frontiers in Plant Science*. 7.
[51] Li, J., Hu, J., Ma, C., Wang, Y., Wu, C., Huang, J., Xing, B. (2016). Uptake, translocation and physiological effects of magnetic iron oxide (γ-Fe_2O_3) nanoparticles in corn (*Zea mays* L.). *Chemosphere* 159: 326–334.
[52] Zhu, H., Han, J., Xiao, J.Q., Jin, Y. (2008). Uptake, translocation, and accumulation of manufactured iron oxide nanoparticles by pumpkin plants. *Journal of Environmental Monitoring* 10(6): 713–717.
[53] Hong, F., Yang, F., Liu, C., Gao, Q., Wan, Z., Gu, F., Wu, C., Ma, Z., Zhou, J., Yang, P. (2005). Influences of nano-TiO_2 on the chloroplast aging of spinach under light. *Biological Trace Element Research* 104(3): 249–260.
[54] Hong, F., Zhou, J., Liu, C., Yang, F., Wu, C., Zheng, L., Yang, P. (2005). Effect of nano-TiO_2 on photochemical reaction of chloroplasts of spinach. *Biological Trace Element Research* 105 (1–3): 269–279.
[55] Zheng, L., Hong, F., Lu, S., Liu, C. (2005). Effect of nano-TiO_2 on strength of naturally aged seeds and growth of spinach. *Biological Trace Element Research* 104(1): 83–91.
[56] Yang, F., Hong, F., You, W., Liu, C., Gao, F., Wu, C., Yang, P. (2006). Influence of nano-anatase TiO_2 on the nitrogen metabolism of growing spinach. *Biological Trace Element Research* 110 (2): 179–190.
[57] Gao, F., Hong, F., Liu, C., Zheng, L., Su, M., Wu, X., Yang, F., Wu, C., Yang, P. (2006). Mechanism of nano-anatase TiO_2 on promoting photosynthetic carbon reaction of spinach. *Biological Trace Element Research* 111(1–3): 239–253.
[58] Gao, F., Liu, C., Qu, C., Zheng, L., Yang, F., Su, M., Hong, F. (2008). Was improvement of spinach growth by nano-TiO_2 treatment related to the changes of Rubisco activase? *Biometals* 21(2): 211–217.
[59] Mingyu, S., Xiao, W., Chao, L., Chunxiang, Q., Xiaoqing, L., Liang, C., Hao, H., Fashui, H. (2007). Promotion of energy transfer and oxygen evolution in spinach photosystem II by nano-anatase TiO_2. *Biological Trace Element Research* 119(2): 183–192.
[60] Yang, F., Liu, C., Gao, F., Su, M., Wu, X., Zheng, L., Hong, F., Yang, P. (2007). The improvement of spinach growth by nano-anatase TiO_2 treatment is related to nitrogen photoreduction. *Biological Trace Element Research* 119(1): 77–88.
[61] Linglan, M., Chao, L., Chunxiang, Q., Sitao, Y., Jie, L., Fengqing, G., Fashui, H. (2008). Rubisco activase mRNA expression in spinach: Modulation by nanoanatase treatment. *Biological Trace Element Research* 122(2): 168–178.
[62] Song, G., Gao, Y., Wu, H., Hou, W., Zhang, C., Ma, H. (2012). Physiological effect of anatase TiO_2 nanoparticles on Lemna minor. *Environmental Toxicology and Chemistry* 31(9): 2147–2152.
[63] Feizi, H., Moghaddam, P.R., Shahtahmassebi, N., Fotovat, A. (2012). Impact of bulk and nanosized titanium dioxide (TiO_2) on wheat seed germination and seedling growth. *Biological Trace Element Research* 146(1): 101–106.
[64] Raliya, R., Biswas, P., Tarafdar, J. (2015). TiO_2 nanoparticle biosynthesis and its physiological effect on mung bean (*Vigna radiata* L.). *Biotechnology Reports* 5: 22–26.
[65] R, R. (2012). Appliance of Nanoparticles on Plant System and Associated Rhizospheric Microflora, J. N. Vyas University Jodhpur.
[66] Tarafdar, A., Raliya, R., Wang, W.-N., Biswas, P., Tarafdar, J. (2013). Green synthesis of TiO_2 nanoparticle using *Aspergillus tubingensis*. *Advanced Science, Engineering and Medicine* 5 (9): 943–949.

[67] Kah, M., Kookana, R.S., Gogos, A., Bucheli, T.D. (2018). A critical evaluation of nanopesticides and nanofertilizers against their conventional analogues. *Nature Nanotechnology* 13(8): 677.
[68] Alshaal, T., El-Ramady, H. (2017). Foliar application: From plant nutrition to biofortification. *The Environment, Biodiversity & Soil Security* 1: 71–83.
[69] Ramesh Raliya, C.F., Chavalmane, S., Nair, R. (2016). Nathan Reed and Pratim Biswas, quantitative understanding of nanoparticle uptake in watermelon plants. *Frontiers in Plant Science* 7: 1288.
[70] Kaphle, A., Navya, P., Umapathi, A., Daima, H.K. (2018). Nanomaterials for agriculture, food and environment: Applications, toxicity and regulation. *Environmental Chemistry Letters* 16 (1): 43–58.
[71] Lemire, J.A., Harrison, J.J., Turner, R.J. (2013). Antimicrobial activity of metals: Mechanisms, molecular targets and applications. *Nature Reviews Microbiology* 11(6): 371.
[72] Mukherjee, A., Majumdar, S., Servin, A.D., Pagano, L., Dhankher, O.P., White, J.C. (2016). Carbon nanomaterials in agriculture: A critical review. *Frontiers in Plant Science* 7: 172.
[73] Tîlmaciu, C.-M., Morris, M.C. (2015). Carbon nanotube biosensors. *Frontiers in Chemistry* 3: 59.
[74] Karny, A., Zinger, A., Kajal, A., Shainsky-Roitman, J., Schroeder, A. (2018). Therapeutic nanoparticles penetrate leaves and deliver nutrients to agricultural crops. *Scientific Reports* 8 (1): 7589.
[75] Fernández, V., Sotiropoulos, T., Brown, P.H. (2013). Foliar Fertilization: Scientific Principles and Field Practices, International Fertilizer Industry Association.
[76] Fageria, N.K., Filho, M.P.B., Moreira, A., Guimarães, C.M. (2009). Foliar fertilization of crop plants. *Journal of Plant Nutrition* 32(6): 1044–1064.
[77] Fernandez, V., Brown, P.H. (2013). From plant surface to plant metabolism: The uncertain fate of foliar-applied nutrients. *Frontiers in Plant Science* 4: 289.
[78] Nel, A., Xia, T., Mädler, L., Li, N. (2006). Toxic potential of materials at the nanolevel. *science* 311(5761): 622–627.
[79] Maynard, A.D., Kuempel, E.D. (2005). Airborne nanostructured particles and occupational health. *Journal of Nanoparticle Research* 7(6): 587–614.
[80] Colvin, V.L. (2003). The potential environmental impact of engineered nanomaterials. *Nature Biotechnology* 21(10): 1166–1170.
[81] Biswas, P., Wu, C.-Y. (2005). Nanoparticles and the environment. *Journal of the Air & Waste Management Association* 55(6): 708–746.
[82] Jiang, J., Oberdörster, G., Elder, A., Gelein, R., Mercer, P., Biswas, P. (2008). Does nanoparticle activity depend upon size and crystal phase?. *Nanotoxicology* 2(1): 33–42.
[83] Oberdörster, G., Stone, V., Donaldson, K. (2007). Toxicology of nanoparticles: A historical perspective. *Nanotoxicology* 1(1): 2–25.
[84] Maynard, A.D., Aitken, R.J., Butz, T., Colvin, V., Donaldson, K., Oberdörster, G., Philbert, M.A., Ryan, J., Seaton, A., Stone, V. (2006). Safe handling of nanotechnology. *Nature* 444(7117): 267.
[85] Jain, A., Ranjan, S., Dasgupta, N., Ramalingam, C. (2016). Nanomaterials in food and agriculture: An overview on their safety concerns and regulatory issues. *Critical Reviews in Food science and Nutrition* (just-accepted): 00–00.
[86] Huang, S., Wang, L., Liu, L., Hou, Y., Li, L. (2015). Nanotechnology in agriculture, livestock, and aquaculture in China. A Review. *Agronomy for Sustainable Development* 35(2): 369–400.

[87] Eichert, T., Goldbach, H.E. (2008). Equivalent pore radii of hydrophilic foliar uptake routes in stomatous and astomatous leaf surfaces–further evidence for a stomatal pathway. *Physiologia Plantarum* 132(4): 491–502.
[88] Delfani, M., Baradarn Firouzabadi, M., Farrokhi, N., Makarian, H. (2014). Some physiological responses of black-eyed pea to iron and magnesium nanofertilizers. *Communications in Soil Science and Plant Analysis* 45(4): 530–540.
[89] Raliya, R., Biswas, P. (2015). Environmentally benign bio-inspired synthesis of Au nanoparticles, their self-assembly and agglomeration. *RSC Advances* 5(52): 42081–42087.

Pratim Biswas and Sukrant Dhawan

Chapter 7 Airborne transmission of SARS-CoV-2: variants and effect of vaccines

7.1 Introduction

The highly infectious SARS-CoV-2 novel coronavirus has resulted in a global pandemic. More than 200 million people are already impacted with infected numbers expected to go up [1], primarily due to evolution of variants of the original virus. The virus can be transmitted via respiratory droplets and aerosols, which can carry infectious viruses, generated during coughing, sneezing, and even talking [2–4]. Since the pandemic began in December 2019, thousands of variants of SARS-CoV-2 have emerged, and the virus has mutated enough to change both its ability to spread through the population and its ability to infect people [5, 6]. Multiple variants of SARS-CoV-2 have been found, of which a few are considered variants of concern (VOCs), given their impact on public health due to their enhanced transmissibility, reduction in neutralization by antibodies obtained through natural infection or vaccination, or the ability to evade detection. As of June 22, 2021, four SARS-CoV-2 VOCs have been identified since the beginning of the pandemic: alpha, beta, delta, and gamma [7]. The delta variant is reported to be nearly twice as contagious as previous strains and may cause even more severe disease among those who are unvaccinated [8]. The viral load of those infected with delta is also reported to be over 1,000 times higher than in those infected with the original form of SARS-CoV-2 [9]. Over the course of a few months, the delta variant has become the predominant strain in the United States, accounting for the vast majority of COVID-19 cases at the end of July 2021 [8].

COVID-19 vaccination is a critical preventive measure to help end this pandemic. People who are fully vaccinated against COVID-19 are highly protected against severe infection, hospitalization, and death caused by the virus [10–12]. The vaccination also reduces the risk of people spreading the virus that causes SARS-CoV-2 [10–12]. Despite the extraordinary speed of vaccine development against COVID-19 and continued mass vaccination efforts across the world, the large pool of people still remains unvaccinated. Low vaccination coverage in many communities is driving the current rapid and large surge in cases associated with the delta variant, and the uncontrolled spread of virus among the unvaccinated people gives the virus more opportunity to spread and mutate into new variants [13]. The emergence of these new

Pratim Biswas, Sukrant Dhawan, Department of Chemical, Environmental and Materials Engineering, University of Miami, Miami, FL 33146

https://doi.org/10.1515/9783110729481-007

variant strains of SARS-CoV-2 threatens to overturn the significant progress made so far in halting the spread of SARS-CoV-2. As the level of vaccination is built worldwide, prevention strategies including masking indoors in public places and in high-density areas must be followed to limit the spread of this epidemic.

It is critical to understand how the transmission of SARS-CoV-2 virus takes place, in order to understand the impact of vaccination and new variants on the risk of transmission and accordingly develop effective public health and infection prevention and control measures. Dhawan et al. [4] developed a comprehensive model based on aerosol dynamics of the respiratory droplets and combined it with the respiratory deposition and a dose–response model that took into account detailed virus considerations, to accurately evaluate the risk of airborne transmission of SARS-CoV-2. The developed model was used to evaluate the risk of infection from coughing, sneezing, and speaking, as a function of different parameters such as viral load, infectivity factor, emission sources, physical separation, exposure time, ambient air velocity, and mask usage [4]. The new infectious variants of SARS-CoV -2 have high infectivity, and the viral load of those infected with delta variants have been reported to be high [8, 9]. The usage of vaccines increases the immunity, and antibodies are developed in the body against the virus [10–12]. Also, viral load is reported to be low when the infected person is infected [14]. Since the developed model [4] accounts for emitted viral load as well as the infectivity of the virus, it was used here to evaluate the relative risk of infection for new infectious variants of SARS-CoV-2 as well as the efficacy of vaccines.

7.2 Droplet lifetime and the effect of relative humidity

The comprehensive methodology to determine the risk of airborne transmission of SARS-CoV-2 is detailed in Dhawan et al. [4]. When an infected individual breathes, speaks, coughs, or sneezes, respiratory droplets containing the virus are emitted. Then the emitted droplets travel through air, and eventually settle on the ground due to gravity. The emitted droplets also undergo evaporation resulting in decrease in their size due to water loss, which increases the time they remain suspended. The distance traversed by these droplets depend on their time of suspension (settling time) or lifetime, which depends on the droplet size. The droplets also undergo evaporation as soon as they are expired. It is important to understand the effect of evaporation on the lifetime of droplets to accurately estimate the droplet transport process. Figure 7.1 outlines a comparison of the evaporation time and the settling time for droplets at the ambient humidity of 50% and temperature of 298 K. The evaporation time of the droplets was calculated by solving the aerosol evaporation equation till the final diameter of the droplets stopped changing due to attaining equilibrium with the surrounding ambient. The settling time of the nonevaporating droplet was evaluated by dividing the height from which droplets (typical value of $H \sim 1.75$ m) are emitted to the settling velocity of the droplets and is given by

$$t_{\text{lifetime}} = \frac{18H\mu}{\rho_{\text{p}} d_{\text{eff}}^2 gC}, \text{ where } d_{\text{eff}} = \alpha d_{\text{p}} \tag{7.1}$$

where g is acceleration due to gravity, μ is air viscosity, d_{p} is droplet diameter, d_{eff} is an effective diameter used to estimate the effective settling velocity depending on the ambient relative humidity (RH, see discussion later), and C is Cunningham's slip correction factor, which accounts for noncontinuum effects while calculating the drag on small particles.

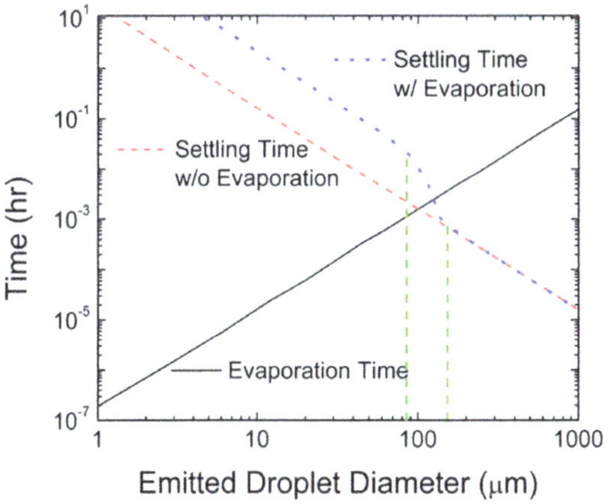

Figure 7.1: Representative evaporation time and settling time calculated for droplets of different initial sizes.

As shown in Figure 7.1, for any particle larger than ~150 μm, the settling time of evaporating droplets is same as the settling time of nonevaporating droplet with same initial diameter ($\alpha = 1$). This is because for particles larger than 150 μm, the settling time is a few orders of magnitude lower than the evaporation time, and the droplets settle before any significant evaporation has taken place. Thus, for droplets larger than 150 μm, d_{eff} is same as emitted diameter d_{p} and α is 1. For particles smaller than 90 μm, the settling time of an evaporating droplet is higher than the settling time of a nonevaporating droplet with the same initial diameter. However, the settling time of an evaporating droplet has similar dependence on the emitted droplet diameter as the settling time of a nonevaporating droplet. This is because for droplets smaller than 90 μm, evaporation takes place very fast as compared to settling. The droplets reach the equilibrium diameter very quickly and settle at that diameter for a relatively longer time. Thus, the settling time of an evaporating droplet of some initial diameter smaller than 90 µm is same as the settling time of nonevaporating droplet with equilibrium diameter. For the respiratory droplets under evaporation at 298 K and 50% RH, the equilibrium diameter is 0.14 times the initial diameter. Thus, for droplets smaller than 90 µm, d_{eff} is 0.14 times emitted diameter d_p or α is 0.14. There is a transition region between 90 and 150 µm. For simplification, 120 µm can be chosen as a transition point between these two regimes, such that for $d_p > 120\,\mu\text{m}$, α is 1 and for $d_p < 120\,\mu\text{m}$ α is 0.14.

The calculations for Figure 7.1 were done at 50% RH, which is typical for indoor environment, but the RH can vary greatly in outdoors. The RH impacts the equilibrium diameter of the evaporating droplets (d_{eff} or α in eq. (7.1)). If the humidity is high, the evaporating droplet will reach equilibrium faster, and the equilibrium diameter of the evaporated droplet will be higher as compared to the equilibrium diameter at lower humidity. Therefore, as RH increases, α increases, and the settling time decreases for small droplet sizes. To generalize, α was evaluated as a function of RH, and the following expression was obtained:

$$\alpha = 0.06773\ \exp(0.01476\text{RH}) + 2.249 \times 10^{-16} \exp(0.3568\text{RH}) \tag{7.2}$$

The accurate lifetime as a function of RH can be calculated by determining α using eq. (7.2), and then substituting it into eq. (7.1). This generalized equation can be readily used as long as there is some knowledge of the ambient RH. It can also be seen from Figure 7.1 that lifetime of droplets >150 µm is ≪1 s. Therefore, it could be concluded that the emitted droplets >150 µm settle quickly are not significant for airborne transmission.

7.3 Risk of airborne transmission

The airborne droplets can be eventually deposited on different surfaces near the infected individual depending on the room layout and the airflow within the room. These droplets carrying infectious virus can also potentially be inhaled by a person in the vicinity. The inhaled droplets with infectious viruses then deposit in the respiratory region of the exposed person, potentially resulting in an infection. The procedure to evaluate the transport of respiratory aerosols and their subsequent deposition in the respiratory region of the exposed individual is described in detail in Dhawan et al. [4]. After the number of infectious viruses deposited in the respiratory region was evaluated, a dose–response model was used to estimate the probability of getting infected [4]. A limitation of conducting risk assessment of SARS-CoV-2 is the lack of quantitative dose–response information for this particular coronavirus. Therefore, in this work, dose–response was approximated from the data available for similar viruses. Existing dose–response models are available for SARS-CoV-1. Watanabe et al. [15] reported an exponential dose–response relationship for SARS-CoV-1. According to this model, if the total expected number of pathogens deposited in the visitor's alveolar region is μ, then the actual integer number of pathogens deposited would follow a Poisson probability distribution. Therefore, if the deposition of just one pathogen can initiate infection, the risk of infection "R" is given by the expression:

$$R = 1 - \exp(-\mu) \tag{7.3}$$

However, not every virus will result in infection. Therefore, to account for actual infectious dose, eq. (7.3) can be modified as follows:

$$R = 1 - \exp(-\sigma\mu) \tag{7.4}$$

where the factor "σ" is the infectivity factor which is the inverse of the number of viruses that can initiate an infection. It is representative of the infectious dose. This factor can be used to describe different variants of the virus which may be more infective. The factor μ represents the total infectious viruses deposited in the respiratory system and is given by

$$\mu = V_{\text{emitted}} \times F \tag{7.5}$$

Thus, the equation for the risk of infection can also be written as follows:

$$R = 1 - \exp((-\sigma \times V_{\text{emitted}} \times F)) \tag{7.6}$$

where V_{emitted} are total viruses emitted, and the factor "F" depends on RH as it affects droplet settling, ventilation rate, background wind velocity, and distance from the infected person. "F" can be calculated using the aerosol transport and respiratory deposition model as presented in Dhawan et al. [4].

"F" was evaluated as a function of background wind velocity (from the infected person to the exposed person), and the distance from the infected person using the detailed transport model presented in Dhawan et al. [4]. The regression equations were used to get the expression of "F" for coughing as a function of wind velocity ($0\ \text{m/s} < u_W < 0.5\ \text{m/s}$ typical indoor values), and the separation from the infected person ($1.8\ \text{m} < L < 7.2\ \text{m}$). The regression equations are given by

$$F = \frac{0.0800 \pm 0.00974}{\left(1 + \left(\frac{L}{5.862 \pm 0.585}\right)^{3.366 \pm 1.406}\right)\left(1 + \left(\frac{u_W}{0.3141 \pm 0.0802}\right)^{0.6993 \pm 0.4018}\right)} \tag{7.7}$$

All these regression equations are accurate at a 95% confidence level. The regression equations have the adjusted $R^2 > 0.97$. Thus, for the case of coughing, the risk of the infection is given by the following equation:

$$R = 1 - \exp\left(\left(-\sigma \times V_{\text{emitted}} \times \frac{0.0800 \pm 0.00974}{\left(1 + \left(\frac{L}{5.862 \pm 0.585}\right)^{3.366 \pm 1.406}\right)\left(1 + \left(\frac{u_W}{0.3141 \pm 0.0802}\right)^{0.6993 \pm 0.4018}\right)}\right)\right) \tag{7.8}$$

The product $\sigma \times V_{\text{emitted}}$ is referred to as the infectious quanta emission rate (Q_{emitted}) in the epidemiological literature [16]. Thus, the Q_{emitted} can be expressed as follows:

$$Q_{\text{emitted}} = \sigma \times V_{\text{emitted}} = \sigma \times V_{\text{load}} \times \text{Vol}_{\text{em, airborne}} \tag{7.9}$$

where V_{load} is the viral load, and $\text{Vol}_{\text{em, airborne}}$ is the volume of airborne droplets emitted during respiratory activities. The viral load is variable among different people and in the same patient during the disease. It lies between 10^3 and 10^{11} copies/mL of the saliva. The infectivity factor σ depends on the variant of virus, vaccination status, and so on, and lies between 0 and 1. The size distribution of emitted droplets during different respiratory activities is discussed in detail in the supplementary information of Dhawan et al. [4]. To determine the $\text{Vol}_{\text{em, airborne}}$, the total volume of droplets below 150 μm emitted during different respiratory activities was evaluated. The cutoff of 150 μm is chosen as the droplets larger than 150 μm settle in ≪1 s (see Figure 7.1) and hence do not contribute to the aerosol transmission.

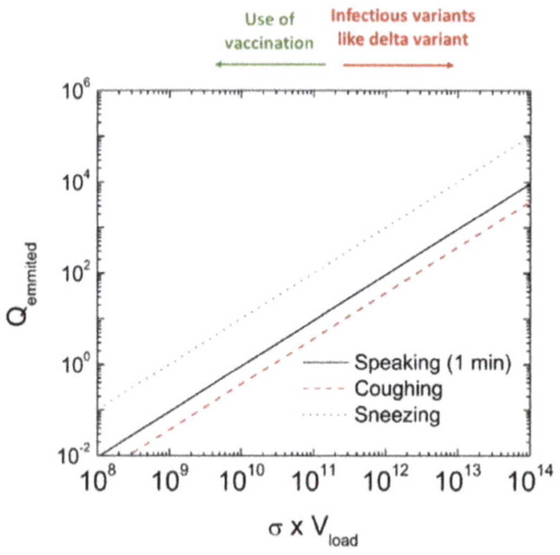

Figure 7.2: The number of quanta emitted from a single sneeze (·········), cough (– – –), and speaking for 1 min (–) from the infected individual as a function of the product of infectivity factor and viral load.

The effect of different variants of SARS-CoV-2 on the risk of transmission and the efficacy of vaccines could be determined by understanding their impact on the infectious quanta emission rate. The different variants of the SARS-CoV-2 virus and the vaccination against the virus impact the Q_{emitted} from the infected person as they impact the virus infectivity and the total viral load in the droplets emitted by the infected person. SARS-CoV-2 variants like delta variant are more infectious and have high infectivity factor "σ," whereas the omicron variant has appropriate values. Also, V_{load} has been reported to be high for people infected with delta variant due to more severe illness. Thus, $Q_{emitted}$ is high for people infected with highly infectious variants. Conversely, if the exposed person is vaccinated, the infectivity factor "σ" decreases due to development of antibodies against the virus. The viral load has been reported to be low when the infected person is vaccinated. Thus, $Q_{emitted}$ is low when either the infected person or the person exposed is vaccinated. The dependence of $Q_{emitted}$ on σV_{load} is shown in Figure 7.2. It can be seen that $Q_{emitted}$ increases linearly with $\sigma \times V_{load}$ and the y-intercept in the log–log plot (Figure 7.2) represents the $\mathrm{Vol}_{em,\ airborne}$ which is different for sneezing, coughing, and speaking.

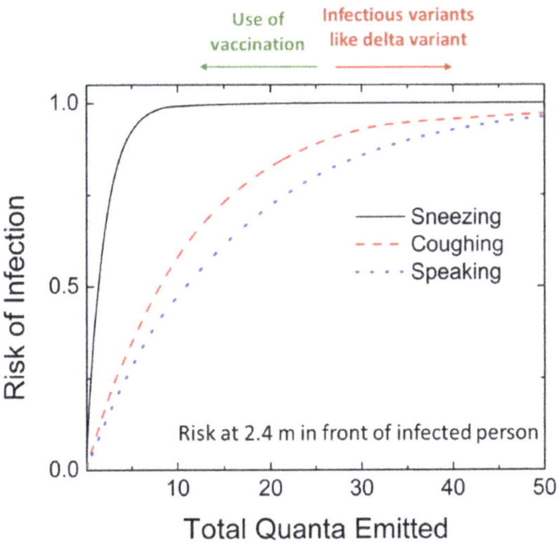

Figure 7.3: Risk of infection from a single sneeze (–), cough (– – –), and speaking for 1 min (·········) from the infected individual, at a separation of 2.4 m as a function of total quanta emitted in the exhaled droplets, in quiescent ambient.

Next, the effect of $Q_{emitted}$ on the risk of transmission of SARS-CoV-2 was evaluated. The ambient temperature and RH were assumed to be 298 K and 50%, respectively, as representative parameters for the results in this work. To determine the effect of total quanta emissions "$Q_{emitted}$" on the risk of infection, eq. (7.8) was used and the risk of infection (when exposed to long time or max. risk) from droplets released during coughing, sneezing, and speaking (for 1 min) was evaluated at a distance of 2.4 m in front of the infected individual in the quiescent ambient air (zero background velocity) as a worst-case scenario. It can be seen from Figure 7.3 that the risk rises from zero, at lower $Q_{emitted}$, increases and eventually saturates to 1 at higher $Q_{emitted}$ values. $Q_{emitted}$ is variable among different people and also in the same patient during the course of the disease and also depends on the variant of SARS-CoV-2 as well as the status of vaccination. The risk of infection is higher for the more infectious variants like delta variants since they result in higher quanta emissions, and the risk of infection decreases on getting vaccinated as it reduces the total quanta emissions (Figures 7.2 and 7.3). It can be seen from Figure 7.3 that the risk is highest for sneezing, followed by coughing, and then by speaking for the same number of quanta emissions. This is because, even though the quanta emissions are same, the average size of droplets emitted during sneezing is smaller as compared to coughing and speaking. Thus, the effect of settling is lowest for sneezing as compared to coughing and speaking, and hence the higher risk. The size distribution of emitted droplets during different respiratory activities is discussed in detail in the supplementary information of Dhawan et al. [4].

7.4 Effect of different variants and vaccination status on risk of infection

Table 7.1: Risk of infection at 2.4 directly in front of the infected individual due to different emission sources for (A) base case (wild-type SARS-CoV-2), (B) infectious delta variant case, and (C) vaccinated individual case.

Emission source	Risk of infection at 2.4 m
A. Base case ($V_{load} = 10^6$ #/mL, $\sigma = 0.1$)	
Sneezing	100%
Coughing	26%
Speaking (1 min)	43%
B. Infectious delta variant case (assumption: 5 times increase in quanta emissions)	
Sneezing	100%
Coughing	78%
Speaking (1 min)	94%
C. Vaccinated case (assumption: 100 times decrease in quanta emissions)	
Sneezing	38%
Coughing	0.3%
Speaking (1 min)	0.5%

Next, to compare the risk of infection due to highly infectious variants and the reduction in risk due to the usage of vaccines, the risk of infection was determined for different cases. There are three cases that were considered in this study: (a) risk for base case, for the wild-type SARS-CoV-2; (b) risk for infectious variants like delta variants; and (c) risk for vaccinated individuals. For the base case, the viral concentration in the saliva was assumed to be 10^6 infectious copies per mL [4]. The infectivity factor, "σ," is not known for SARS-CoV-2. For SARS-CoV-1, this "σ" lies in the range of 0.01–0.10 [15]. Since SARS-CoV-2 is highly infectious compared to SARS-CoV-1, "σ" was assumed to be 0.10 for the base cases in our simulations. Thus, for the base case, Q_{emitted} was calculated by assuming $\sigma = 0.1$, and viral concentration to be 10^6 infectious copies per mL of saliva. It can be seen from Table 7.1 that for the base case, the risk of infection at 2.4 m distance is 26% from a cough, 100% from sneeze, and 43% from speaking for 1 min.

It should be noted that different variants of the virus such as delta variants have different infectivity rates, that is, higher "σ." An individual infected with delta variants also have higher viral load [9]. Thus, Q_{emitted} is higher for delta variant of SARS-CoV-2 as compared to the base case (Figure 7.2). It can be seen from Table 7.1 that for 5 times increase in Q_{emitted}, the risk of infection increases and is 78% from a cough, 100% from sneeze, and 94% from speaking for 1 min. The increase in the risk of transmission is more than 180% for coughing and 118% for speaking (1 min) with 5 times increase in Q_{emitted}. Thus, highly infectious variants of SARS-CoV-2 can possess significantly higher risk as compared to the base case.

Next, the effect of vaccination on SARS-CoV-2 transmission was examined. Upon vaccination, the body's immunity increases because antibodies that can counter the virus are developed. Therefore, more virus exposure is needed to initiate the infection, which implies that there is a decrease in "σ" when the exposed individual is vaccinated. Also, if the infected individual is vaccinated, the viral load is reported to be very low. Thus, Q_{emitted} is significantly lower upon vaccination as compared to the base case (Figure 7.2). It can be seen from Table 7.1 that for 100 times decrease in Q_{emitted}, the risk of infection decreases and is less than 1% from a cough, 38% from sneeze, and less than 1% from speaking for 1 min. COVID-19 vaccines are a critical tool in overcoming this pandemic as they significantly reduce the risk of transmission. All of the current vaccines are highly effective in preventing hospitalization and death which also helps to relieve the strain on the healthcare system. As the percentage of vaccinated people increases, the transmission of the virus will slow down. It will result in fewer opportunities for the virus to mutate, which can help prevent the emergence of any other variants. People who are fully vaccinated are highly protected against severe infection, hospitalization, and death caused by the virus. However, coronavirus cases among the fully vaccinated are still being seen. Therefore, prevention strategies such as masking indoors in public places and in high-density areas must be followed to limit the spread of this epidemic till the significant percentage of the population gets vaccinated. It should be noted that values for the infectivity factor (hence Q_{emitted}) used in this work are approximate since the data are not yet available for variation factors due to different variants of SARS-CoV -2 and the impact of vaccination. Clearly, as more representative values are obtained by experimentation, later, the appropriate values can be used.

In summary, a comprehensive model has been developed to account for the transmission of viruses. Using appropriate emission rates and infectivity factors, risks of infection can be estimated. Such models can then be used as effective tools for public health protection. For example, the use of masks, the impact of effective physical distancing, and the role of vaccination can be quantitatively estimated to demonstrate enhanced protection to prevent the spread of infectious diseases such as COVID-19.

References

[1] Worldometer Website. https://www.worldometers.info/coronavirus/. (08/10/2021).

[2] Klompas, M., Baker, M.A., Rhee, C. (2020). Airborne Transmission of SARS-CoV-2: Theoretical Considerations and Available Evidence, Jama.

[3] Morawska, L., Cao, J. (2020). Airborne transmission of SARS-CoV-2: The world should face the reality. *Environment International* 139: 105730.

[4] Dhawan, S., Biswas, P. (2021). Aerosol Dynamics Model for Estimating the Risk from Short-Range Airborne Transmission and Inhalation of Expiratory Droplets of SARS-CoV-2, Environmental Science & Technology.

[5] Phan, T. (2020). Genetic diversity and evolution of SARS-CoV-2. *Infection, Genetics and Evolution* 81: 104260.

[6] Alam, I., Radovanovic, A., Incitti, R., Kamau, A.A., Alarawi, M., Azhar, E.I., Gojobori, T. (2021). CovMT: An interactive SARS-CoV-2 mutation tracker, with a focus on critical variants. *The Lancet Infectious Diseases* 21(5): 602.

[7] Lauring, A.S., Hodcroft, E.B. (2021). Genetic variants of SARS-CoV-2 – what do they mean? *Jama* 325(6): 529–531.

[8] CDC Website. https://www.cdc.gov/coronavirus/2019-ncov/variants/delta-variant.html. (08/10/2021)

[9] Li, B., Deng, A., Li, K., Hu, Y., Li, Z., Xiong, Q., Liu, Z., Guo, Q., Zou, L., Zhang, H., Lu, J. (2021). Viral infection and transmission in a large well-traced outbreak caused by the Delta SARS-CoV-2 variant. medRxiv. DOI: https://doi.org/10.1101/2021.07.07.21260122.

[10] Vahidy, F.S., Pischel, L., Tano, M.E., Pan, A.P., Boom, M.L., Sostman, H.D. ... Omer, S.B. (2021). Real world effectiveness of COVID-19 mRNA vaccines against hospitalizations and deaths in the United States. medRxiv. DOI: https://doi.org/10.1101/2021.04.21.21255873.

[11] Thompson, M.G., Burgess, J.L., Naleway, A.L., Tyner, H.L., Yoon, S.K., Meece, J., Olsho, L.E., Caban-Martinez, A.J., Fowlkes, A., Lutrick, K., Gaglani, M. (2021). Interim estimates of vaccine effectiveness of BNT162b2 and mRNA-1273 COVID-19 vaccines in preventing SARS-CoV-2 infection among health care personnel, first responders, and other essential and frontline workers – eight US locations, December 2020–March 2021. *Morbidity and Mortality Weekly Report* 70(13): 495.

[12] Andrejko, K., Pry, J., Myers, J.F., Jewell, N.P., Openshaw, J., Watt, J., Jain, S., Lewnard, J.A. (2021). Early evidence of COVID-19 vaccine effectiveness within the general population of California. medRxiv. DOI: https://doi.org/10.1101/2021.04.08.21255135.

[13] Niesen, M., Anand, P., Silvert, E., Suratekar, R., Pawlowski, C., Ghosh, P., Lenehan, P., Hughes, T., Zemmour, D., OHoro, J.C., Soundararajan, V. (2021). COVID-19 vaccines dampen genomic diversity of SARS-CoV-2: Unvaccinated patients exhibit more antigenic mutational variance. medRxiv. DOI: https://doi.org/10.1101/2021.07.01.21259833.

[14] Levine-Tiefenbrun, M., Yelin, I., Katz, R., Herzel, E., Golan, Z., Schreiber, L. ...Kishony, R. (2021). Initial report of decreased SARS-CoV-2 viral load after inoculation with the BNT162b2 vaccine. *Nature Medicine* 27(5): 790–792.
[15] Watanabe, T., Bartrand, T.A., Weir, M.H., Omura, T., Haas, C.N. (2010). Development of a dose-response model for SARS coronavirus. *Risk Analysis: An International Journal* 30(7): 1129–1138.
[16] Sze To, G.N., Chao, C.Y.H. (2010). Review and comparison between the Wells–Riley and dose-response approaches to risk assessment of infectious respiratory diseases. *Indoor Air* 20(1): 2–16.

Index

https://doi.org/10.1515/9783110729481-008

Lightning Source UK Ltd.
Milton Keynes UK
UKHW051021111122
411897UK00007B/27

9 783110 730968